KARLSRUHER BERICHTE
ZUR INGENIEURBIOLOGIE

Heft 41

Degradation of Whey in an Anaerobic Fixed Bed (AnFB) Reactor

Marisa Handajani

universitätsverlag karlsruhe

Dissertation, Universität Karlsruhe (TH), Fakultät für Bauingenieur-, Geo- und Umweltwissenschaften, 2004

Impressum

Universitätsverlag Karlsruhe
c/o Universitätsbibliothek
Straße am Forum 2
D-76131 Karlsruhe

www.uvka.de

Print on Demand

ISSN 0172-8709
ISBN 3-937300-12-0

Acknowledgements

I especially thank Prof. Dr. rer. nat. J. Winter, who guided and worked with me in all phases of this dissertation and whose wide interests created a stimulating environment in which to work. I sincerely thank Prof. Dr.-Ing E.h. H. H. Hahn, PhD who gave his time and large knowledge and helped to bring this dissertation to successful end.

I sincerely acknowledge the financial support of DFG-Graduiertenkolleg III GK-GRK 366/2-00: „Grenzflächenphänomene in aquatischen Systemen und wässrigen Phasen". I would like also to thank Prof. Dr. Dr. habil. F. H. Frimmel as the representative of Graduiertenkolleg 366 and Dr. B. Gordalla as the program coordinator for their supports during my participation in this project.

I also want to acknowledge all colleges of the Institut für Ingenieurbiologie und Biotechnologie des Abwassers for their support, care and friendship. Many thanks go to my family and friends who cared and encouraged me and lighted my life.

M. Handajani

Abstract

An Anaerobic Fixed Bed (AnFB) reactor was run as an upflow anaerobic reactor with an arrangement of supporting material for growth of a biofilm. The supporting material was made from Liapor-clay-polyethylene sinter lamellas (Herding Co., Amberg). The AnFB reactor was used for treating high concentrations of whey-containing wastewater. Optimal operating conditions for whey treatment at a concentration of COD in the influent of around 50 g whey$\cdot l^{-1}$ were found for a hydraulic retention time (HRT) in the range of 4-8 days or an organic loading rate (OLR) less than 10 kg COD$\cdot m^{-3}\cdot d^{-1}$. This is a higher load than normally applied in praxis reactors.

Accumulation of volatile fatty acids (VFAs) happened when the AnFB was supplied with surplus whey solution at a high OLR or when it was oxygenated. VFAs were accumulated faster when the HRT was changed from 12 days to 6 days compared to a change of HRT from 6 days to 4 days. However, at a HRT of 6 days, the accumulated VFAs were completely degraded after an adaptation period of about 5 days, whereas the accumulated VFAs at a HRT of 4 days remained constant upon time and could not be degraded during further incubation.

The conversion process (acetogenesis and methanogenesis) of VFAs was influenced by the pH in the reactor. Acetate and n-Butyrate were converted faster at neutral or slightly alkaline pH, while propionate was degraded faster at slightly acidic pH-value.

The population in the AnFB contained hydrogen-utilizing methanogenic bacteria, formate-utilizing methanogenic bacteria, methanol-utilizing methanogenic bacteria, acetoclastic methanogenic bacteria and sulfate-reducing bacteria as the final-stage organism of whey degradation. Acetogenic and methanogenic bacteria grew slower and were present at much lower numbers than acidogenic bacteria. This made the acid degradation rate less than the acid production rate. The minimal HRT in the whey reactor was thus dependent on acid degradation rates.

Acetate-utilizing methanogens seemed to be unable to grow as single cells. They preferred to grow in a particulate or attached manner on a support material. The biofilm on the support materials provided a lower redox potential and an anaerobic environment that was obligately needed by these bacteria. The addition of a reducing agent was necessary to keep the few culturing acetoclastic methanogens in suspended cultures active.

H_2/CO_2 was the best methanogenic substrate for the bacteria in the effluent suspension of whey reactor, followed by formate and methanol. The least degradable substrate in suspension cultures was acetate. The optimal H_2 gas concentration for methanogens was provided at 2.25 bar.

Ferric ions addition or the addition of a mix of minerals improved acetate degradation and methane production rates more than two-folds. The redox potential $\pm$ reducing agent was low enough for methanogenesis.

An AnFB-reactor would be a suitable means for stabilizing wastewater from dairy processing. Liapor-clay-polyethylene sinter lamellas in a regularly arrangement could be the substratum for biofilm formation. A minimum HRT of 4-6 days should be planned or a maximum OLR rate 10 kg $COD \cdot m^{-3} \cdot d^{-1}$ not exceeded.

Contents

List of Tables

List of Figures

Chapter 1

INTRODUCTION

1.1 Introduction

The world annual liquid whey production for 1974 was estimated to be 74 million tons, 95% of which was cheese whey. In the United States about 1 million tons of solid whey were produced in 1974, which is equivalent to approximately 17 million tons of liquid whey. For comparison, the production in 1970 in the United States was only 59% of the total production of cheese and casein whey in Western Europe (Kosaric and Asher, 1985).

Whey is a by product of cheese and casein manufactures. It has a high biological oxygen demand which is caused by its protein and carbohydrate content and it creates a problem when disposed as wastewater. The worldwide production of fluid whey by the cheese and casein industries runs into millions of tons, and yet effective utilization of this material is not well developed.

When disposing whey as a wastewater into a sewage treatment plant, it has been estimated that 50 kg whey were equivalent to the waste produced by 22 people every day. In other words, a cheese plant producing 50.000 kg of whey per day requires a treatment plant of about the same size like a city with a population of 22.000 inhabitant equivalents (Gillies, 1974).

Traditionally whey was returned to the farmer as animal feed or spreaded on the fields. According to Shahani and Mathur (1971), only about 56% of the whey solids were utilized for human food and animal feed. Today, since large cheese industries are common and world cheese production is steadily by increasing this mode of whey utilization is not desirable. The recovery process for milk sugar, protein and salts from whey is carried out by evaporation and drying. It is a fractional proposition that misses about one-fifth of the production, which escapes as rinse water. This water characteristically has several times the strength of domestic sewage and it has been reported to induce problems in the sewers and in domestic activated sludge plants, when present at a level as low as 11% of the total BOD used. Parenthetically the consensus is that the treatment of dilute whey is facilitated by the dilution and the improved nutrient balance in mixtures with domestic sewage.

Wastewaters from dairy industries are characterized by high putrescibility, high oxygen demand, and the production of poor settling sludges following biological treatment. Wastewater containing whey from the cheese manufacture is notorious for causing waste treatment problems, whether treated alone or in conjunction with other wastes, including domestic sewage. In addition to possessing high putrescibility, whey also creates problems of pH control and nutrient deficiencies during biological treatment (Gillies, 1974).

Anaerobic operation has recently been accepted as an effective mean of treating high strength wastewaters. For wastewater with a total BOD in excess of 4000 $mg \cdot l^{-1}$ an anaerobic contact process was cheaper than an aerobic process. There were many applications which were described for wastewater from the meat packing, brewing, pharmaceutical, chemical and food processing industries (Rüffer and Rosenwinkel, 1991).

There are some benefits of anaerobic processes in treating organic wastewater, which can be summarized as follows:

(1) Low cellular yields. This process requires only relatively few inorganic nutrients for treating wastewater and produces only little amounts of surplus sludge.

(2) The major gas product is methane, a valuable and sustainable energy resource. The second major gas in carbon dioxide.

(3) No requirement of oxygen in the process, in contrary strictly anaerobic conditions and the absence of nitrate are required.

(4) Good capability for degrading a variety of natural organic compounds. Produce a residual sludge which is inoffensive and useful as a soil conditioner and low grade fertilizer.

(5) Anaerobic digestion of primary sewage sludge effectively reduces the concentration of pathogenic microorganisms.

Beside the benefits, there are also some disadvantages of anaerobic processes, which are:

(1) The major microbial population is inherently unstable, thereby requiring that the system must be kept under close supervision and control.

(2) The maximum specific growth rates of the organisms are relatively low. That makes their response to perturbations sluggish and requires the system to be designed with relatively long MCRT´s, even if the process is operated at an elevated temperature.

(3) The application of this process or alternatively of an aerobic trickling filter process is limited since many synthetic organic chemicals are resistant to anaerobic (or aerobic) degradation.

(4) Require a heat exchanger that makes the capital cost of anaerobic reactors high. However, lower capital cost for aerobic treatment system is opposed by much higher operational cost than for anaerobic system.

The benefits of an anaerobic process can be applied for treating wastewater that contains whey. A suitable reactor for this type of wastewater is an anaerobic fixed bed reactor because (1) it has a valuable end product, in this case methane, and (2) a high degree of waste stabilization may be accomplished while producing only small amounts of excess biomass, even though this process needs a longer time for degrading the waste and it requires a constant temperature in the mesophilic or thermophilic range for an effective stabilization of the wastewater.

1.2 Objective

This research orientation was the biological treatment of a whey-containing wastewater using an anaerobic process. The study was carried out in laboratory scale with an anaerobic fixed bed reactor, using a synthetic wastewater, prepared from whey powder. The main focuses were investigating the feasibility of a biological process for treating highly concentrated carbohydrate-containing wastewater under anaerobic conditions and also the capability of the anaerobic population for degradation of intermediately excreted volatile fatty acids, which were produced during the phases of high loading or overloading in this process. The specific objectives of this research were:

(1) to observe the performance of an anaerobic fixed bed reactor in treating whey containing wastewater under changing loading conditions.

(2) to investigate the parameters which cause a reactor failure.

(3) to establish and optimize the degradation of whey containing wastewater in an anaerobic fixed bed reactor.

(4) to determine the kinetic parameters of degradation of whey-containing wastewater.

1.3 Theoretical Background and Literature Review

1.3.1 Biological Processes: Aerobic versus Anaerobic Wastewater Treatment

A biological process for wastewater treatment is an operation in which biochemical transformations are catalysed by living organisms, either in the presence or absence of oxygen. The purification is associated with the energy metabolism of aerobic or anaerobic microorganisms. The applications of biological processes are very diverse: they can be found in the food and beverage industry, the pharmaceutical industry, the chemical industry, the metal industry and also in domestic wastewater or sewage sludge treatment. The biological anaerobic process is directed to remove soluble organic matter as biogas, to degrade insoluble organic matter by hydrolysis to soluble organic matter and to convert soluble inorganic matter by redox reactions in conjunction with e.g. precipitation.

There are two major environments for biological processes, aerobic and anaerobic ecosystems. An aerobic environment provides dissolved oxygen in sufficient quantity so that oxygen will not be a limiting parameter in the process. In this environment, oxygen acts as the terminal electron acceptor in microbial metabolism and growth of the microbes occurs in an efficient manner. In anaerobic environment, dissolved oxygen is either not presents or its concentration is too low so that oxygen availability limits or completely prevents an aerobic metabolism. In anaerobic environment some substances other than oxygen serve as the terminal electron acceptors. If the substance is an organic substrate, the system is called fermentation, while when the electron acceptor is an inorganic substance, the culture is said to undergo anaerobic respiration (Grady and Lim, 1980). Sulfate reduction would be an example for "anaerobic respiration".

In treating wastewater, there is a practical rule for determining the environment which is suitable for the process. Aerobic treatment processes are particularly suitable for removing organic matter in the concentration range between 50 and 4000 $mg \cdot l^{-1}$, measured as biodegrable chemical oxygen demand (COD). For COD concentrations above 4000 to 50.000 $mg \cdot l^{-1}$, an anaerobic treatment or at least pre-treatment is frequently used to reduce the quantity of oxygen which must be provided in subsequent aerobic final cleaning operations (Grady and Lim, 1980). The experiments and application results in the last thirty years, showed that an anaerobic treatment was preferably used for COD concentrations equal or higher than 1500 $mg \cdot l^{-1}$. Beyond the lower limit of COD concentrations for aerobic biological processes, physical and chemical processes are more efficient. Perspectives of anaerobic digestion were recently summarized by Ahring (2003).

1.3.1.1 Anaerobic Processes

Before insoluble organics can be consumed, they must be solubilized, in order to facilitate transport across the cell membrane. The reactions are usually hydrolytic processes and are catalysed by enzymes which have been released to the medium by aerobic and anaerobic bacteria. The soluble molecules of hydrolysis of polymer are used as carbon and energy sources by anaerobic bacteria, which carry out their fermentation. The oxidized end products of those fermentations are primarily short-chain volatile fatty acids, such as acetic, propionic, butyric, valeric and caproic acid. This process is referred to as acidogenesis and the responsible organisms are acid-producing "fermentative" bacteria. The reduced end products of fermentation depend upon the nature of the culture and the environmental conditions in the reactor.

There is a group of acetic acid-producing bacteria within the acidogenic bacteria which has special enzymes that allow them to oxidize reduced coenzymes without passing the electrons on to an organic acceptor, thereby releasing hydrogen gas (H_2) to the medium. Except for these hydrogen producing fermentative bacteria, acetogenic bacteria utilize volatile fatty acids larger than acetic acid to produce acetic acid and H_2 from even-numbered fatty acids and acetic acid, carbon dioxide (CO_2) and H_2 from odd numbered fatty acids. This is, however, only possible if the hydrogen partial pressure is below 10^{-4} atm. for propionate conversion or 10^{-5} atm. for butyrate conversion. The group of these bacteria is called obligately hydrogen-producing bacteria and their activity is known as acetogenesis or hydrogenogenesis. While there is an unclear distinction between only acid-producing and acid plus hydrogen-producing bacteria, these two groups of bacteria are generally referred to as non-methanogenic bacteria and their integrated metabolism with hydrogen-consuming methanogens or sulfate reducer results primarily in formic acid, acetic acid, CO_2 and H_2.

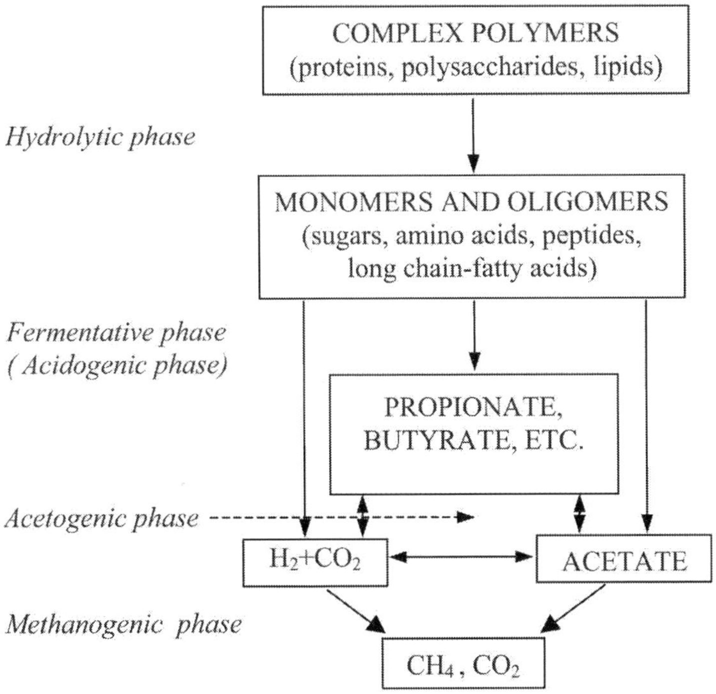

Figure 1 Multi step scheme for the flow of carbon in the complete anaerobic conversion of organic matter to methane.

When no hydrogen is formed during the process, the nonmethanogenic phase results in an insignificant reduction of COD because all electrons released in the oxidation of organic compounds are passed to organic acceptors, which remain in the medium as energy-conserving metabolites. On the other hand, when H_2 is formed, it represents a gaseous product which escapes from the medium, thereby causing a reduction in the energy content, and thus the COD of the liquid is decreasing.

The next step in the anaerobic process is the methanogenic phase. In anaerobic processes, the products of the non-methanogenic phases are utilized by methanogenic bacteria to produce methane gas. Some methane-producing bacteria have the ability to use other volatile fermentation products, such as primary and secondary or cyclic alcohol, as hydrogen donor for methane production. One mole of methane requires two moles of oxygen to oxidize it to CO_2 and water. This means that for each 16 grams of methane produced and lost to the atmosphere that equals to the removal of 64 grams of COD from the liquid. At standard temperature and pressure, it corresponds to 0.34 m^3 of methane or 0.68 m^3 of biogas for each kg of COD.

Prescott and Dunn (1949) define fermentation as a process in which chemical changes are brought about in an organic substrate. Whether these are carbohydrates or protein or fat or some other types of organic material is not relevant. Biochemical catalysts known as 'enzymes' and produced by specific types of living microorganisms are required for conversion of organic substances.

While according to Elsden (1952), fermentation may be defined as a biological process in which chemical energy is made available for growth by oxidative reactions, the ultimate hydrogen acceptors are substances other than oxygen.

In milk or milk product, lactic acid bacteria are the prime "agents" of fermentation. Whey as one of the waste products in dairy industry contains lactic acid and lactic acid bacteria, when it is fermented. The lactic acid bacteria of the genera Streptococcus, Lactococcus, Leuconostoc and Lactobacillus are Gram positive bacteria. They can be classified into cocci and rods. Some of them are mesophiles which grow fastest at 20-30°C and others are moderate thermophiles, growing best at 35-45 °C. Lactic acid bacteria are nonspore-forming, non motile and almost catalase-free microorganisms.

Lactose contributes about 70% of the solids in whey, and expectably plays a dominant role in determining the properties of whey. The fermentation of whey-containing wastewater involves sugar and protein metabolism. This metabolism includes the transport of sugar into the cell and its further breakdown as follows:

1. Uptake of sugar into the bacterial cell and formation of hexose monophosphates. During growth, lactose must be transported across the cytoplasmic membrane (Figure 2).
2. Further metabolism will vary with the species considered and results in various end products.

Pyruvate is a key intermediate. This compound is the precursor for all lactose fermentation pathways as presented in Figure 2.

1. The Homofermentative Lactic Acid Fermentation

   Sugar is metabolized via the glycolytic or Embden-Mayerhof pathway: galactose-6-P enters by the tagatose pathway. One disaccharide molecule (e.g. lactose) yields two hexose molecules (glucose + galactose). Two molecules of glyceraldehydes-3-P are finally formed from one hexose molecule. Figure 2 shows that lactose is fermented according to equation 1:

   Equ. 1: Lactose + 4 H_3PO_4 + 4 ADP → 4 lactic acid + 4 ATP + 3 H_2O

2. The Heterofermentative Lactic Acid Fermentation

   Heterofermentative lactic acid bacteria cannot ferment sugar via the glycolytic pathway. The presence of glucose-6-P-dehydrogenase and phosphoketolase permits metabolism by the pentose phosphate pathway. Phosphoketolase hydrolyses 6-P gluconate to CO_2 and a pentose-5-P, which is converted to glyceraldehydes-3-P and acetyl-P. The conversion of glyceraldehydes-3-P to lactic acid proceeds like in the glycolytic pathway. Under these conditions, lactose is fermented according to equation 2:

   Equ. 2:
   Lactose + 2 H_3PO_4 + 2 ADP → 2 lactic acid + 2 ethanol + 2 CO_2 + 2 ATP + H_2O

3. Alternative Metabolic Pathways

Besides homofermentative and heterofermentative lactic acid fermentation, other end products may be formed if NADH can be oxidized in another way (Figure 2)

a. Via the pyruvate-formate lyase (PFL) reaction

This enzyme is essential for the anaerobic pyruvate metabolism leading to formate and acetyl-Co-A. The later component is further metabolized into ethanol and acetate. Strictly anaerobic conditions are required because PFL is rapidly inactivated if O_2 is present.

b. Via the action of NADH oxidase and NADH peroxidase in the presence of O_2.

Most of the lactic acid bacteria applied in dairy industry are microaerophilic, which means that the bacteria can grow at low oxygen pressure.

Pyruvate oxidase / NADH peroxidase has been found in some lactic acid bacteria. These enzymes catalyze the conversion of pyruvate to acetyl-P, which is hydrolyzed into acetic acid and ATP (equation 3):

Equ. 3:

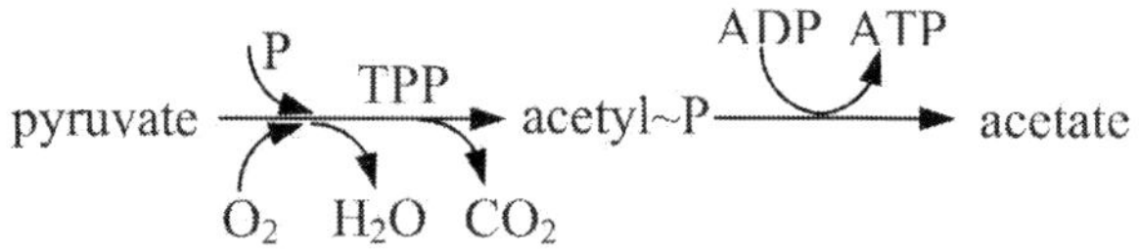

TPP : Thiamine pyrophosphate

c. An important alternative metabolic pathway is the citrate metabolism. Citrate is not used as an energy source but is metabolized only in the presence of fermentable sugars like lactose. Additionally, pyruvate is formed during citrate metabolism so that more of it becomes available than is required for oxidation of NADH. Specific end products then include acetic acid, CO_2 and C-4 products (diacetyl) (equation 4):

Equ. 4:

$$COOH \bullet CH_2 \bullet C(OH)COOH \bullet CH_2 \bullet COOH \rightarrow$$

$$CH_3 \bullet COOH + CO_2 + CH_3 \bullet CO \bullet COOH \rightarrow \rightarrow \text{diacetyl (Figure 3)}$$

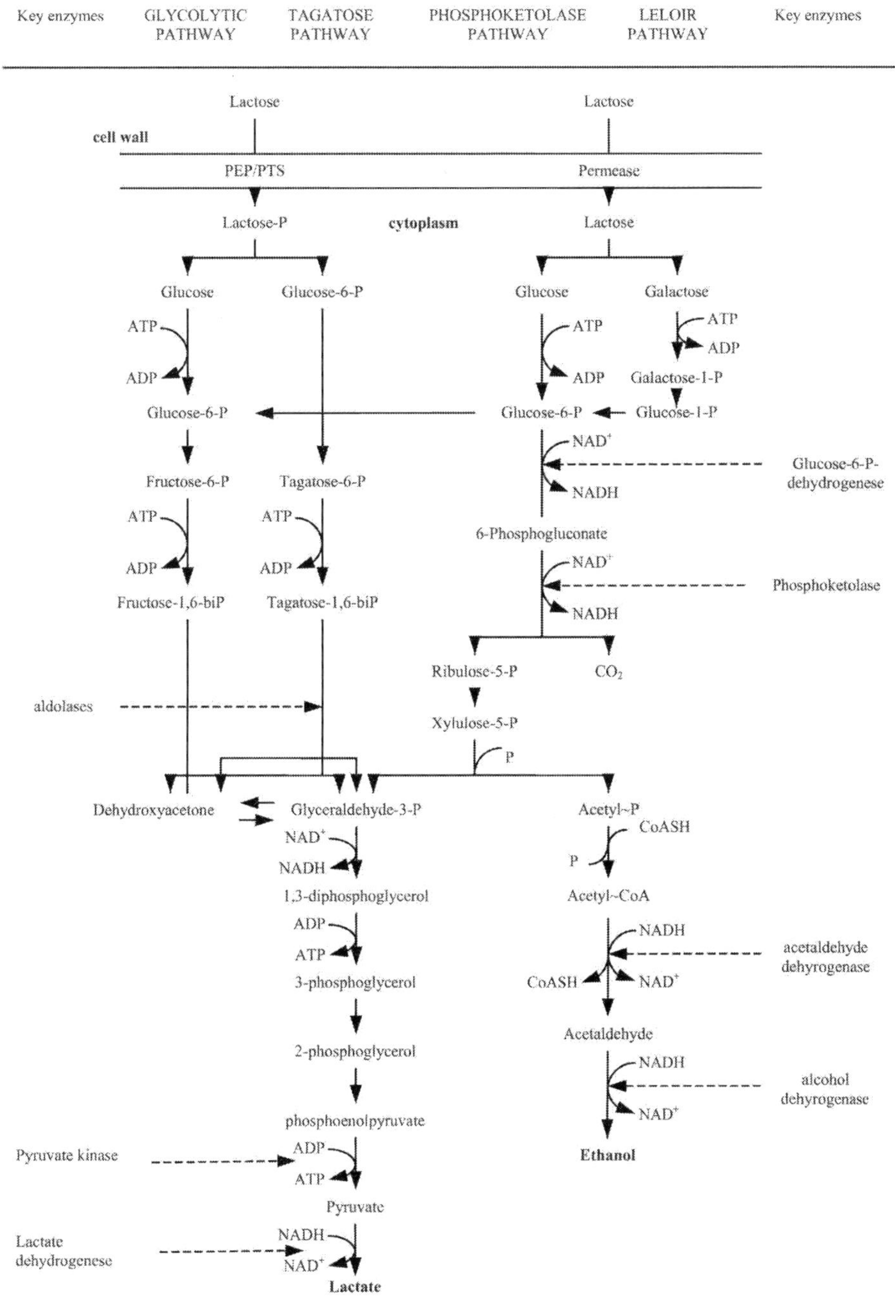

Figure 2 Metabolism of lactose in lactic acid bacteria. (Walstra et al., 1999 - modified).

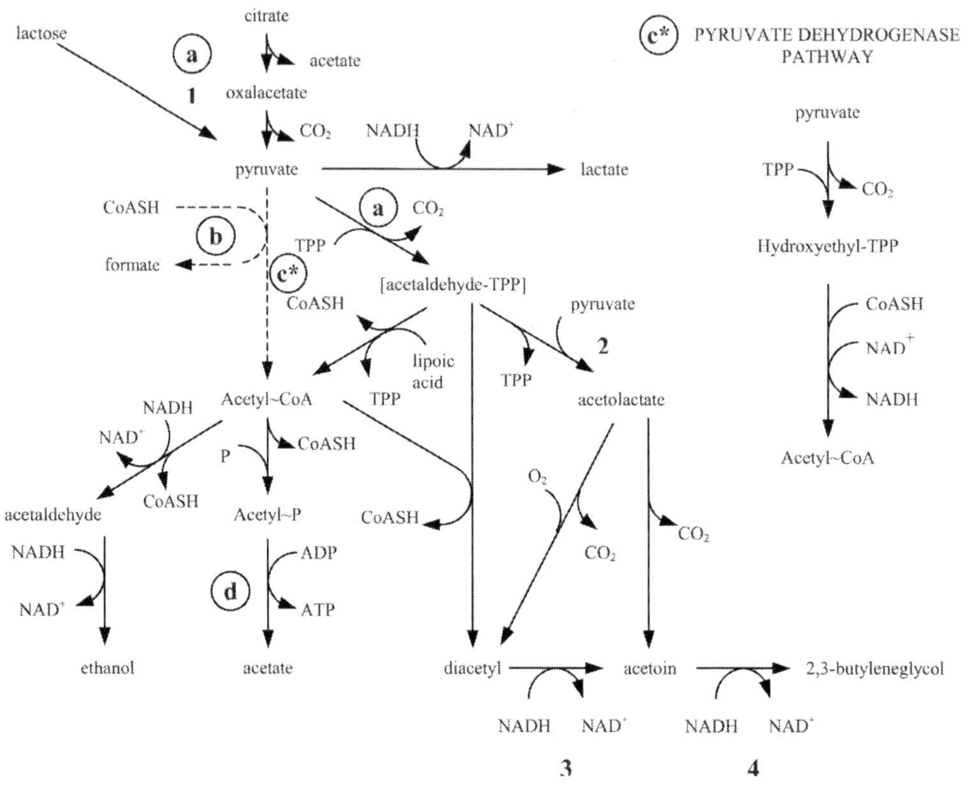

Figure 3 Pyruvate metabolism in lactic acid bacteria. The common metabolic pathways result in production of lactic acid or lactic acid plus ethanol, as is shown in **Figure 2**. Important alternative pathways are: a. Citrate metabolism. Enzymes involved include citrate lyase (1), acetolactate synthase (2), diacetyl reductase (3), and acetoin reductase (4). b. Pyruvate-formate lyase pathway. c. Pyruvate dehydrogenase pathway. d. Also of importance for heterofermentative bacteria in the presence of other hydrogen acceptors: acetic acid is produced rather than ethanol. TPP, thiamine pyrophosphate. (After various sources.) (Walstra et al., 1999 - modified).

d. Production of Acetaldehyde

Acetaldehyde is predominantly accumulated by lactic acid bacteria that have no alcohol dehydrogenase enzyme. These bacteria therefore cannot reduce acetylaldehyde (formed via the pyruvate-lyase pathway) to ethanol.

4. The Propionic Acid Fermentation

The propionic acid bacteria are Gram positive, nonspore-forming, non motile rods. They grow under anaerobic and microaerophilic conditions. They ferment lactic acid, carbohydrates and polyalcohols with the formation of propionic acid, acetic acid and CO_2. They are usually strongly catalase positive and contain a cytochrome system. The stoichiometry of propionic acid fermentation with lactate as a substrate is shown in equation 5:

Equ. 5: $$3\ CH_3{\bullet}CHOH{\bullet}COOH \rightarrow 2\ CH_3{\bullet}CH_2{\bullet}COOH + CH_3{\bullet}COOH + CO_2$$

5. The Formation of Acetic and Carbonic Acids

In the fermentation of lactate, the constancy of the ratio between the amounts of acetate and carbonic acids points to the existence of a common precursor, pyruvic acid. It is formed as a result of oxidation of lactic acid (equation 6):

Equ. 6: $$CH_3{\bullet}CHOH{\bullet}COOH \leftrightarrows CH_3{\bullet}CO{\bullet}COOH + 2\ \langle H \rangle$$

With the formation of pyruvic acid, the conditions now exist for its decarboxylation coupled with the reduction of oxalacetic acid to succinic acid (equation 7 and 8):

Equ. 7: $$2CH_3{\bullet}CO{\bullet}COOH + 2H_2O \leftrightarrows 2CH_3{\bullet}COOH + 2CO_2 + 4\langle H \rangle$$

Equ. 8: $$H{\bullet}CO{\bullet}CH_2{\bullet}COOH + 4H \overset{-H_2O}{\leftrightarrows} COOH{\bullet}(CH_2)_2{\bullet}COOH$$

Thus an accounting has been made for two end products of fermentation, acetic and carbonic acid (succinate). Propionic acid is derived from succinic acid to the extent that succinic acid is decarboxylated (equation 9):

Equ.9: Succinate + propionyl-CoA $\leftrightarrows$ succinyl-CoA + propionate
$\hookrightarrow \ \longrightarrow$ ....

6. The Formic Acid Fermentation

Formic acid production is a characteristic of e.g. *E. coli.* Mixed cultures produce formic acid from pyruvic acid (equation 10):

Equ.10: $CH_3{\bullet}CO{\bullet}COOH + H_2O \rightarrow CH_3{\bullet}COOH + HCOOH$

7. The Formation of H_2

The conversion of formic acid to H_2 and CO_2 by formate dehydrogenase is reversible (equation 11):

Equ. 11: $HCOOH \leftrightarrows H_2 + CO_2$

Hydrogenlyase is an alternative way to produce H_2 in the absence of an intermediate electron carrier.

8. The Butyric Acid Fermentation

Microorganisms which produce butyric acid belong either to the genus Clostridium or Eubacterium. They are Gram positive, spore-forming (clostridia) or non-spore forming (*Eubacterium sp.*) rods. Both are strictly anaerobic. For the formation of butyric acid, C-3 compounds may be utilized, such as glycerol, pyruvate and lactate. It would appear that butyric acid formation is not a direct result of splitting of a hexose into C-4 and C-2 components. Butyrate is formed from 2 acetyl-CoA units instead (see reaction sequences, equation 12-17).

The mechanism postulated for conversion of acetate to butyrate also explains that the conversion of lactate to butyrate by *C. lactoacetophilum* requires acetate. One mole of added acetate is consumed per mole of butyric acid formed, while *Butyribacterium rettgeri* requires CO_2 as a precursor of additional acetate in the formation of butyric acid from lactic acid.

The fermentation of butyric and the conversion of pyruvate to acetate by *C. acetobutylicum* are inhibited by CO. Thus an iron containing protein complex appears to be involved (Levinton and Marth, 1965).

The pathway of butyrate fermentation (equation 12-17):

Equ. 12:

$$2\,CH_3{\bullet}CO{\bullet}COO^- + 2\,CoAS \underset{+2\,CO_2}{\overset{-2\,CO_2}{\rightleftharpoons}} \underset{\text{Acetyl-CoA}}{2CH_3{\bullet}CO{\bullet}SCoA} \underset{+\,CoASH}{\overset{-\,CoASH}{\rightleftharpoons}} \underset{\text{Acetoacetyl-CoA}}{CH_3{\bullet}CO{\bullet}CH_2{\bullet}CO\text{-}SCoA}$$

Equ. 13:

$$\underset{\text{Acetoacetyl-CoA}}{CH_3{\bullet}CO{\bullet}CH_2{\bullet}CO\text{-}CoA\ SCoA} \underset{-2H}{\overset{+2H}{\rightleftharpoons}} \underset{\text{Vinylacetyl-CoA}}{CH_2{=}CH{\bullet}CH_2{\bullet}CO\text{-}}$$

Equ. 14:

$$\underset{\text{Acetoacetyl-CoA}}{CH_3{\bullet}CO{\bullet}CH_2{\bullet}CO\text{-}CoA} \underset{-2H}{\overset{+2H}{\rightleftharpoons}} \underset{\text{Crotonyl-CoA}}{CH_3{\bullet}CH{=}CH{\bullet}CO\text{-}SCoA} + H_2O$$

Equ. 15:

$$\text{Vinylacetyl-CoA} \underset{-2H}{\overset{+2H}{\rightleftharpoons}} \underset{\text{Butyryl-CoA}}{CH_3{\bullet}CH_2{\bullet}CH_2{\bullet}CO\text{-}SCoA}$$

Equ. 16:

$$\text{Crotonyl-CoA} \underset{-2H}{\overset{+2H}{\rightleftharpoons}} \underset{\text{Butyryl-CoA}}{CH_3{\bullet}CH_2{\bullet}CH_2{\bullet}CO\text{-}SCoA}$$

Equ. 17:

$$\text{Butyryl-CoA} \rightleftharpoons \text{butyrate} + CoAS$$

1.3.1.2 Non-methanogenic Bacteria

The non-methanogenic bacteria consist of various groups of facultative and obligately anaerobic bacteria, with obligate anaerobes as the predominant population. This population mostly carries out the important hydrolytic and fermentative reactions. The major end product of the non-methanogenic bacteria are the short chain volatile fatty acids (with acetic, propionic, and butyric as the most important products). The concentrations of various products are determined by the environmental conditions and the specific growth rates imposed upon the culture. These shift the predominant species within the population and cause internal changes within individual organisms.

The fermentative, hydrogen-producing bacteria are those with the most important internal changes. When these species do not produce hydrogen, they must use organic compounds as acceptors for the electrons removed during biological oxidation. It results in the formation of reduced products such as, butanol, lactic acid and succinic acid in addition to an oxidized product such as acetate. On the contrary, when the hydrogenase enzyme system is kept active, few reduced organic compounds need be formed and acetate becomes the major end product. Acetogenic hydrogen-forming bacteria can use the reduced organic compounds produced by the fermentative bacteria and convert them to acetate as well. This situation maximizes the production of acetate while the production of reduced end products is minimized.

Pure cultures of non-methanogenic hydrolytic and fermentative bacteria in general have growth rate (μ_{max}) around 1 h^{-1}, where as the growth rates of acetogenic bacteria are in the order of several days, e.g. 0.1-0.3 d^{-1}.

1.3.1.3 Methanogenic Bacteria

There are two distinct groups of methanogenic bacteria. One group provides its energy from oxidation of molecular hydrogen while the other oxidizes acetate.

(1) Hydrogen-oxidizing methanogens.

All methanogenic bacteria which oxidize hydrogen are strictly anaerobes. They provide their energy from hydrogen oxidation and their carbon from carbon dioxide. With this autothrophic mode of life, the cell yield of this group is low. Reactions involved in energy methabolism and cell growth are summarized below (Figure 4):

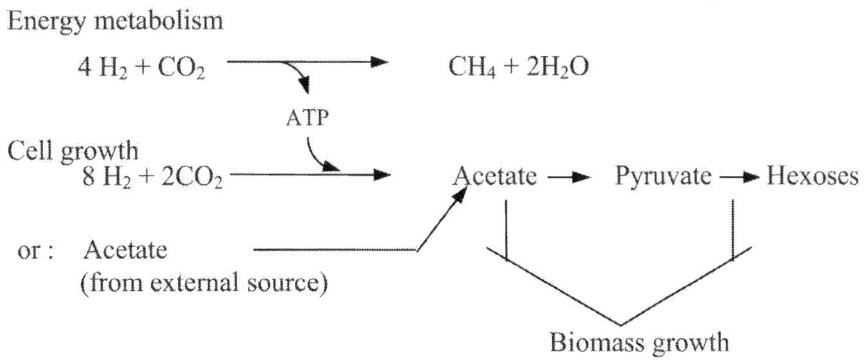

Figure 4 Simplified scheme of the energy and cell growth metabolism of hydrogenolytic methanogens.

These bacteria are very sensitive to pH changes and are inhibited by pH values outside the range 6.7 – 8.4. At a high partial pressure of hydrogen with a pH around neutral, this group of bacteria can grow with maximal growth rates, μ_{max}, around 0.04 h^{-1}.

Most of the bacteria in this group cannot use acetate as an energy substrate but this does not mean that they cannot use it to form cell mass. A few methanogens can use acetate as a terminal electron acceptor with the following reaction (equation 18):

Equ. 18: $$CH_3COOH + 4\ H_2 \rightarrow 2\ CH_4 + 2\ H_2O$$

With this mechanism, methanogenic bacteria may prevent the accumulation of hydrogen by removing the hydrogen that is produced by non-methanogenic bacteria. Hydrogen accumulation can inhibit the hydrogen-forming bacteria.

(2) Acetate-oxidizing methanogens

This group contain obligate anaerobes but they are relatively more tolerant towards dissolved oxygen. They have growth rates smaller than hydrogen-utilizing methanogens. They usually are unable to utilize hydrogen and carbon dioxide or formate as energy substrates simultaneously to acetate cleavage. This group plays a major role in producing methane by mixed cultures degrading complex substrates. Methane is formed by cleavage of acetate (equation 19):

Equ. 19: $CH_3COOH \rightarrow CH_4 + CO_2$

If carbohydrates are the substrates for methanogenesis, glycolysis delivers 2 mole acetate + 2 mole CO_2 + 8<H>. Subsequently, 2 mole methane are formed from acetate and 1 mole methane from CO_2-reduction. Thus, if carbohydrates are the main source of substrates for methanogenesis, methane and CO_2 are produced in same amounts. However, some CO_2 is solubilized in the water phase, leading to an apparent deficit of biogas and a higher than the theoritical methane proportion.

1.3.1.4 Factors Affecting Performance of Anaerobic Processes

TEMPERATURE. The temperature has an effect on anaerobic fermentation because a temperature incline will increase the specific growth rates of the complex microorganisms to a different extend. This factor becomes important since anaerobic systems depend on the interacting population. Changes of only a few degrees Celsius can cause an imbalance between the major contributions to the population which can lead to a process failure. Decreasing the temperature can cause a decline in μ_{max} and an increase in K_s, requiring longer hydraulic retention times (HRT) at lower temperature to guard against washout of methanogenic bacteria.

pH. The pH of the culture medium will exert an effect upon all individuals of the microbial population. Hydrogen-utilizing methanogenic bacteria in general are more sensitive to pH changes and generally show a relatively narrow range in which growth will occur. Methanogenesis is almost totally inhibited at a pH $\leq$ 6.2. In an anaerobic reactor, the main concern is about methanogenic bacteria, so that the pH of operation should be maintained near 7.0. Severe problems may arise when the pH is dropping below 6.5.

TOXICITY. It is the inhibiting effect of certain concentrations of natural or xenobiotic substances on the specific growth rate of bacteria when the material is present in excess. At the beginning, the increase of the concentration of the material will increase the specific growth rate until the μ_{max} is reached. This phase is called the stimulatory region. A further increase of concentration will have no effect on the specific growth rate. The further increase will reach a threshold value at which the specific growth rate starts to decrease. At this point toxicity is reached and any concentration in excess of that will be toxic to the culture and reduce its viability.

VOLATILE ACIDS. The effects of volatile acids on microorganisms in anaerobic reactors are complicated because the acids can also affect the pH of the medium. Little inhibition by volatile acids will occur at neutral pH. According to Hobson and Summers (1967), neither acetic nor butyric acids have any significant toxic effect upon hydrogen-utilizing methanogenic bacteria at concentrations up to 10.000 $mg \cdot l^{-1}$, when the pH of the medium is kept around neutrality. On the other hand, propionic acid shows partial toxicity to methanogenic bacteria at a concentration of 1000 $mg \cdot l^{-1}$ at neutral pH.

AMMONIA. Ammonia is a material that comes up in wastewater degradation, in this case from protein-containing wastewater. When protein is degraded, the nitrogen is released as ammonia. There are two forms of ammonia which are depended upon the pH of the system: ammonium ions (NH_4^+), or dissolved free ammonia. The most toxic form is free ammonia. Under mesophilic conditions, a concentration of 80 - 150 mg NH_3-N·l^{-1} at pH of 7.5, will inhibit the system. Under thermophilic conditions at pH of 7.2 – 7.3, the inhibitory concentration of free ammonia for acetoclastic methanogens was 250 mg·l^{-1} NH_3-N and for hydrogenolytic methanogens 500 mg·l^{-1} NH_3-N (Gallert and Winter 1997). The toxicity of free ammonia can be prevented by running the reactor at pH $\leq$ 7.2. Below this pH value, ammonia will be in the form of ammonium ions so that a total ammonia concentration approaching 300 mg·l^{-1} can be tolerated with little effect. An indicator of ammonia toxicity is a build up in volatile acids.

LIGHT METAL CATIONS. The toxic effects of light metal cations appear with the implementation of pH-control systems with bases as a titrant. The cations of sodium, potassium, calcium and magnesium are of particular concern. Stimulatory and inhibitory concentrations of these ions are compiled in Table 1.

Table 1 Stimulatory and inhibitory concentration of light metal cations[a].

Cation	**Concentration in mg·l^{-1}**		
	Stimulatory	**Moderately Inhibitory**	**Strongly Inhibitory**
Sodium	100-200	3500-5500	8000
Potasium	200-400	2500-4500	12000
Calcium	100-200	2500-4500	8000
Magnesium	75-150	1000-1500	3000

[a] modified from Grady and Lim (1980)

SULFIDE. In anaerobic reactors, sulfide is produced by reduction of sulfates, which are present in the influent and from degradation of protein. Sulfides in soluble form have toxic effects to bacteria since they precipitate essential trace ions in the cytoplasm. The toxicity of sulfides can be reduced by adding heavy metal ions to the medium since most of them form highly insoluble precipitates with sulfide. Sulfide also can be removed as gaseous hydrogen sulfide. The concentration of soluble sulfide depends upon both, the pH of the liquid and the composition of the gas space. The effect of soluble sulfide on anaerobic digestion is listed in the Table 2.

Table 2 Effect of soluble sulfide in the anaerobic process[a].

Concentration	Unit	Effect
<100	$mg \cdot l^{-1}$	Tolerable, with little or no acclimation
100 – 200	$mg \cdot l^{-1}$	Little effect after acclimatization
>200	$mg \cdot l^{-1}$	Strongly inhibitory

[a] summarised from Grady and Lim (1980)

HEAVY METALS. Although some heavy metal ions are essential for growth of methanogen in trace amounts, many heavy metals are toxic to anaerobic populations at a low concentration (Table 3). The toxic effect and the concentration of soluble heavy metals can be reduced by precipitation with sulfides. The approximate dose is 0.5 $mg \cdot l^{-1}$ of sulfide per 1 $mg \cdot l^{-1}$ of heavy metal ions (Kugelman et al.,1971).

Table 3 Concentration of soluble heavy metals exhibiting 50% inhibition of an aerobic digester[a].

Cation	Approximate Concentraiton in $mg \cdot l^{-1}$
Fe^{++}	1 – 10
Zn^{++}	10^{-4}
Cd^{++}	10^{-7}
Cu^{+}	10^{-12}
Cu^{++}	10^{-16}

[a] modified from Grady and Lim (1980)

RAPIDITY OF CHANGES. The difference in the growth rate between the non-methanogenic and the methanogenic population in anaerobic processes must be considered when any changes will be done in the reactor environment. The changes should be made at rates which can be tolerated by methanogenic bacteria. This is a precaution for preventing a process failure.

HYDRAULIC LOADING. In an anaerobic reactor, loading is defined as the mass of organic matter added per unit volume per unit time. The hydraulic loading is related to the active volume of the reactor, time in the reactor and the concentration of the feed. Hydraulic loading is used as an empirical design parameter. Since there is no relationship between hydraulic loading and mean cell residence time (MCRT), especially in biofilm reactor there is no strict correlation between loading and process performance.

1.3.2 Kinetics of Biochemical Reactions

1.3.2.1 Background

The total rate of a series of biological reactions depends on the catalytic activity of the enzyme those catalyses the bottle-neck reaction. The enzyme kinetics has been defined by Michaelis and Menten for a single reaction which is involving a single substrate. This equation was extended for simulating kinetics of wastewater treatment, which involves various substrates and mixed culture (Benefield und Randal, 1980, Grady and Lim, 1980, Hartmann, 1992).

In general, an enzyme-catalysed reaction involves a reversible reaction of enzyme (E) and substrate (S) to form an enzyme-substrate complex. At the end, the complex decomposes irreversibly or reversible to free enzyme (E) and product (P). The reaction is presented as follows (equation 20):

Equ. 20: $$E + S \underset{K_2}{\overset{K_1}{\rightleftharpoons}} ES \xrightarrow{K_3} E + P$$

K_1, K_2 and K_a are the rate constants for the reactions designated.

When the concentration of the ES-complex is constant, the dynamic steady state condition is established. The ES-complex formation rate is equal to the ES-complex break down rate. For this dynamic equilibrium rate, the Michaelis-Menten equation 20 can be rearranged into equation 21.

Equ. 21: $$v = v_{\max} \frac{S}{K_m + S}$$

v = reaction rate, ($mmol \cdot l^{-1} \cdot t^{-1}$)

v_{max} = maximum reaction rate ($mmol \cdot l^{-1} \cdot t^{-1}$)

S = concentration of subtrate ($g \cdot l^{-1}$ or mM)

K_m = Michaelis constant ($g \cdot l^{-1}$ or mM) saturation constant, that is the substrate concentration when the reaction rate is equal to $v_{max}/2$

K_m = $(K_2 + K_3) / K_1$

The Michaelis-Menten equation illustrates a phenomenon of substrate saturation which is typical for enzyme reactions. The equation shows that the enzyme-catalyzed reaction rate at low concentration of substrate is close to proportional to the substrate concentration. It means that the reaction rate approximately forms a first order kinetics with increasing substrate concentration. Later on, a decreasing curve is reached, when the substrate is further increasing. This is referred to as the region of mixed order. For a still further increase of the concentration of the substrate, the reaction rate is close to a constant high value and it does no longer depend on the substrate concentration. In this region the reaction rate is of zero order with respect to substrate concentration. This observation implies two special cases of the Michaelis-Menten equation.

When the substrate concentration is much greater than K_m, the value of K_m can be ignored in the sum term found in the denominator of the equation. This will reduce the equation 21 to equation 22:

Equ. 22: $$v \cong v_{max}$$

It shows that the reaction rate is constant and equal to its maximum value. Under this situation the reaction follows zero order kinetics which means the reaction is independent of the concentration of substrate.

If the concentration of substrate is much smaller than K_m, [S] can be neglected in the sum term of the denominator of the equation and the Michaelis-Menten equation reduces to equation 23:

Equ. 23: $$v = \frac{v_{max}}{K_m} S$$

The maximum growth rate (v_{max}) and K_m are both constants in a enzyme-substrate-system and the reaction is proportional to the concentration of the substrate. The reaction follows a first order kinetics.

According to Segel (1968), a kinetic can be a zero order reaction when the substrate concentration is equal or higher than 100 times of K_m ($S \geq 100\ K_m$) while for first order reactions the concentration of the substrate is equal or less than one-hundredth K_m ($S \leq 0.01\ K_m$).

The Michaelis-Menten equation represents a continuum for defining enzyme-catalyzed reaction. For example an experiment was started with a large amount of substrate and the kinetics of the reaction was followed over time. When there was no new substrate addition, the reaction followed zero order kinetics initially. With an excess of substrate, the reaction rate was limited by the enzyme capacity. As the substrate became used up, the reaction became a substrate-limited reaction and its followed fractional order reaction. When the concentration of the substrate was very little, the substrate metabolism rate became controlled by its concentration and this resulted in a first order kinetics.

The Michaelis-Menten equation is a general expression for catalytic enzyme processes. A wastewater treatment process which involves high concentrations of microorganisms, can be defined as a generally substrate-limited process. So it can follow a first order kinetics.

1.3.2.2 Transformation of Michaelis-Menten Equation

The kinetics constants (v_{max} and K_m) usually are provided from experimental results. Examining the data reveals that v_m can be estimated by measuring the reaction rate at high substrate concentration. Furthermore, rearrangement of Michaelis-Menten equation reveals that when the substrate concentration is numerically equivalent to K_m, v is one-half of v_{max}. By plotting the data as in the graph of Figure 5, the abscissa value corresponding to an ordinate value of $v_{max}/2$ is equal to K_m. Even though this technique is simple in concept, it is seldom used because it is inaccurate, thereby making it difficult to obtain reliable values for the parameters. A more reliable estimation can be obtained by transforming the Michelis-Menten equation into a linear form.

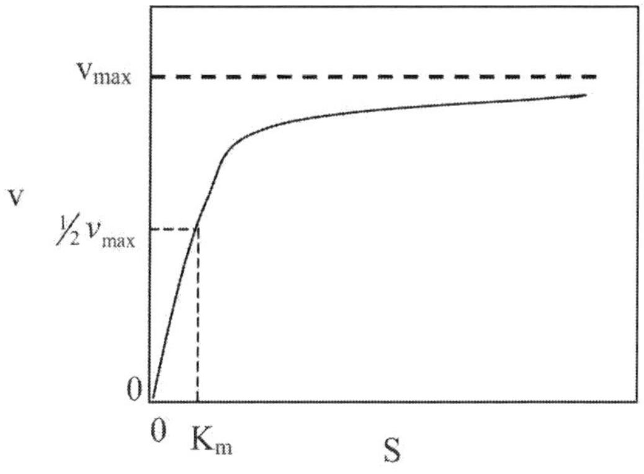

Figure 5 Plot of the "velocity" of an enzymatic reaction as a function of reactant concentration as depicted by the Michaelis-Menten equation (Grady and Lim, 1980).

There are several techniques which can be used for calculating the constant. These techniques are based on mathematic transformation of the Michaelis-Menten equation. The most common used technique is obtained by taking the reciprocal of both sides of the basic equation and rearranging it to equation 24:

Equ. 24:
$$\frac{1}{v} = \frac{1}{v_{max}} + \frac{K_m}{v_{max}}\frac{1}{S}$$

This equation is called the Lineweaver-Burk equation, the v_{max} may be obtained from the intercept on the y-axis and the K_m from the slope of the line. In Lineweaver-Burk plots the most accurately known values (around v_{max}) will tend to cluster near the origin, while those that are least accurately known will cluster far from the origin and thus will have the most influence upon the slope (Grady and Lim, 1980).

The other technique is obtained by multiplying both sides of the Lineweaver-Burk equation by S and rearranging it to equation 25:

Equ. 25:
$$\frac{S}{v} = \frac{S}{v_{max}} + \frac{K_m}{v_{max}}$$

This equation was originally suggested by Hanes (Grady and Lim, 1980) and yielded a straight line when S/v is plotted as a function of S. In this case v_{max} is obtained from the slope and K_m from the intercept. In the Hanes equation, the points near v_{max} are spread out so that the slope, $1/v_{max}$, may be determined accurately. Often, however, the intercept appears near the origin which makes an accurate determination of K_m difficult. In spite of that, this technique can be quite reliable, especially when a least square fit is used.

The basic equation may also be linearized by multiplying both sides by (K_m + S) and dividing both sides by S to give equation 26:

Equ. 26:
$$v = v_{max} - K_m\frac{v}{S}$$

Thus a plot of v versus v/S will give a straight line with a slope equal to $-K_m$ and an intercept equal to v_{max}. Such a plot is called a Eadie-Hofstee plot. The Eadie-Hofstee equation contains the substrate-dependent variable v, in both coordinates, which makes the use of the least squares fitting technique invalid. This technique does spread the data out, however, so that reasonable fits may be obtained by eyes.

Walker developed another technique for determining the v_{max} and K_m with the following equation 27 (Hartmann, 1992):

Equ. 27:

$$\frac{S_0 - S}{t} = -K_m \frac{1}{t} \ln \frac{S_0}{S} + v_{max}$$

S_0 = initial concentration of substrate ($t = 0$, $S = S_0$)

t = time

Plotting $(S_0 - S)/t$ versus $1/t \cdot \ln (S_0/S)$ will give a straight line with a slope equal to $-K_m$ and the intercept with the ordinate equal to v_{max}.

1.3.3 Anaerobic Reactors

A reactor is a "room" where the chemical and biochemical transformations proceed. There are several types of bioreactors. Based on the growth mode of microorganisms, reactors can be classified as slurry reactors and fixed film reactors. In the first reactor type, the microorganisms grow suspended in the fluid, while in the second one, the microorganisms grow as a film on an immobile support material, the substratum. Another classification is based on the feeding system: batch fermentations, fed-batch fermentations and continuous fermentation in the turbidostat or the chemostat mode. Concerning the mixing, there are continuously stirred tank reactor (CSTR) and plug flow reactors.

In application for anaerobic wastewater processes, there are two kinds of CSTR, anaerobic contact reactors and anaerobic digesters. Anaerobic contact reactors are used to remove soluble organic matter under anaerobic condition by sludge recycling to increase the active biomass concentration in the reactor. These reactors are also used to treat waste containing a mixture of soluble and suspended organics. Anaerobic digestion reactors without sludge return are used for stabilizing suspended organic matter at high concentrations and containing a high initial population density.

1.3.3.1 Anaerobic Fixed Bed Reactors

An anaerobic fixed bed reactor is a bioreactor in which the anaerobic organisms grow as a biological film on the surface of a non-moving supporting material. The population of microorganisms in this bioreactor takes the substrate and the nutrients from and releases the metabolite products to the medium that flows over the film. This reactor is classified as an open system since the biomass is retained on the surface of the supporting material until the maximal biomass thickness on the boundary layer is reached. When these conditions are achieved, the biofilm will loose its adhesiveness and it will be washed out through the effluent of the reactor. This system does not need a biomass recirculation.

There are many kinds of anaerobic fixed-bed reactor or sometimes alternatively called anaerobic filter reactors. In contrary to aerobic trickling filters, anaerobic fixed-bed reactors obtain a bottom feed so that the supporting material is completely submerged in the wastewater. By the attached growth mode, the biomass remains a longer time in the system. This will allow also organisms with longer generation time to grow in the biofilm, aside from the fast growing species.

A fixed bed reactor is more suitable for a soluble waste, because it can prevent the loss of biomass that is dispersed in the effluent of completely stirred tank reactors. Attached growth of the wastewater flora prolongs the cell residence time. There are five advantages for applying an anaerobic filter process (Bennefield and Randall, 1980):

(1) Soluble waste can be effectively treated with an anaerobic filter.

(2) No form of recycle is required with anaerobic filters because the microorganisms are growing attended to remain in the filter and are not lost in the effluent.

(3) Because of the high biomass solid concentration maintained in the filter reactor, it is possible to operate at a lower temperature than would normally be possible for other types of anaerobic process.

(4) Very low volumes of sludge are produced by the anaerobic filter.

(5) Start up and shutdowns are easier with an anaerobic filter than with other types of anaerobic processes.

In operating a biological process, it is essential to determine the mean cell residence time (MCRT) within the reactors. The MCRT becomes the primary parameter during design of anaerobic reactor. A higher degree of waste stabilization and a greater production of methane are obtained as the MCRT is increased, although the exact relationship between MCRT and performance must be determined experimentally for each particular wastewater.

A minimum MCRT is required to achieve relatively complete destruction of volatile solids. It depends upon many factors, including the composition and the particle sizes available for degradation. In most anaerobic reactors, a MCRT in excess of 15 days leads to a constant methane production per unit of biodegradable COD added to the reactor. As the MCRT is reduced below 10 days, methane production in CSTR is decreasing in a manner consistent with the washout of methane-forming bacteria.

In another case, at 35°C, a MCRT in excess of 10 days did not produce significant improvement in effluent quality (Grady and Lim, 1980). The decline of methane production was accompanied by an increase of volatile fatty acids in the effluent which were formed by degradation of the original substrate by non-methanogenic bacteria. At a steady state during constant production of methane by the methanogenic population, the fatty acids were not accumulating, but completely degraded. The degradation of volatile solids is related to the MCRT until the minimal value is reached, at which washout of the complex population starts. When the MCRT is reduced below the washout point of methanogenic bacteria, which comprise normally only 10-20% of the total population, the stabilization of wastewater will be terminated because there is no more biogas production (sink of energy) and the system will revert to a simple fermentation with fatty acid accumulation.

At a very small MCRT, which allows no growth of the methanogenic population, most of the carbon in the influent is converted to volatile fatty acids. When an anaerobic process is used for treating solubles in wastewater, the decisive parameters are the rates at which the volatile acids are utilized by the acetogenic and methanogenic population, rather than the rates at which the original substrates are converted to volatile acids. Among the volatile acids, propionic and butyric acid are the most important substrates for acetogenesis and acetate, CO_2 and H_2 for methanogenesis. This shows that volatile fatty acid conversion to acetate, CO_2 and H_2 and subsequently methane formation may become rate limiting in anaerobic operations. Thus, anaerobic fermentation may be considered as a single step process from the kinetic point of view.

1.3.4 Whey

There are two basic disadvantage of whey as a raw material for industrial processes; (a) its perish ability and (b) its low total solid content. Nevertheless the utilization of dried whey powder is highly developed in most of the Western European Countries. The most significant use is in animal feeding.

Whey is the serum of milk resulting after removal of fat and casein. Whey is a by product in cheese and casein production. There are two kinds of whey, depending on the type of milk coagulation used, either sweet or sour whey. Sweet whey is obtained if milk is coagulated by proteolytic enzymes, such as chymosine and pepsine or microbial enzymes produced from *Mucor miehei* and *Mucor pusillus*. It comes from the production of cheddar, Swiss and Italian varieties of cheese. Acid whey is obtained if milk is coagulated by acids mainly by lactate from lactic acid fermentation of lactose. It comes from soft cheese production, with cottage cheese as a predominant product. The chemical composition of whey depends upon the type of coagulation used (Table 4).

It contains roughly half the solid of the whole milk from which it is derived and most of the water soluble vitamins and minerals. A typical whey contains around 6.5% total solids. These include lactose (68–72%), protein (12–13%), minerals (8–9%) plus small amounts of fat and lactic acid. Some examples of whey composition are listed in Table 4.

There is a long list of products from whey solids or whey powder. Whey products are added to beverages, soups, sauces, dressings and creams among others. Whey powder is used as an ingredient of animal feed, since it is a cheap source of high quality protein and carbohydrates.

Table 4 Composition of fluid and dried whey (Kosikowski, 1982).

Component	Fluid Sweet Whey[a] (%)	Fluid Acid Whey[b] (%)	Condensed Acid Whey (%)	Dried Sweet Whey (%)	Dried Acid Whey (%)
Total solids	6.35	6.5	64.0	96.5	96.0
Moisture	93.70	93.50	33.5	3.5	4.0
Fat	0.5	0.04	0.6	0.8	0.6
Protein, total	0.8	0.75	7.6	13.1	12.5
Lactose	4.85	4.90	34.9	75.0	67.4
Ash	0.50	0.80	8.2	7.3	11.8
Lactic Acid[c]	0.05	0.40	12.0	0.2	4.2

[a] From cheddar cheese
[b] From cottage cheese
[c] Estimated true lactic acid after substituting for acidity

Chapter 2

MATERIALS AND METHODS

2.1 Analytical Procedures

2.1.1 Chemical Oxygen Demand (COD)

The Chemical Oxygen Demand (COD) was determined using the method of Wolf and Nordmand (1977). COD is used as a measure of the oxygen equivalents of the organic matter content in waste water samples that are susceptible to oxidation by a strong chemical oxidant. Potassium dichromate in an acidic solution (H_2SO_4 and H_3PO_4 containing Ag_2SO_4 as catalyst) has an oxidizing effect on almost all organic substance and on a number of inorganic compounds or ions.

For analysing, 1 ml liquid sample and 1.5 ml COD reagent were mixed well and heated in a thermoblock at 150°C for 2 hours. After oxidation, the green Cr^{3+} ions were measured spectrophotometrically at 615 nm. The absorbance of the oxidation result was converted to the COD concentration by comparison with a standard curve of glucose (0 – 2000 $mg \cdot l^{-1}$).

2.1.2 Volatile Fatty Acids (VFAs)

The volatile fatty acid composition was determined with a PACKARD model 437 A gas chromatograph equipped with a FID detector and a Teflon column (2 mm i.d. x 2 m length). The temperature settings used were as follows: column at 180°C, injector and detector at 210°C. This equipment used hydrogen gas (30 $ml \cdot min^{-1}$) as a burning gas and required an air flow of approximately ten times the hydrogen gas flow (300 $ml \cdot min^{-1}$) for complete combustion. Nitrogen served as the carrier gas at a flow rate of 30 $ml \cdot min^{-1}$. The Teflon column was filled with chromosorb C101 (80-100 mesh; Sigma, Deisenhofen) and equilibrated (activated) with phosphoric acid.

For measuring the VFA content of samples, the supernatant of centrifuged wastewater samples was acidified 1:1 with 4% H_3PO_4. One μl of acidified solution was injected with micro syringe (Hamilton, USA). As a reference a mixed standard fatty acid concentration was injected. The mixed standard of fatty acids contained acetate, propionate, (iso- and n-) butyrate and (iso- and n-) valerate with a concentration of 50 mM for each acid. The fatty acid concentration in the samples should be less than 5 mM. Analysis conditions were described by Gallert and Winter (1997).

2.1.3 Biogas Composition

The biogas produced during anaerobic digestion of whey was measured with a gasometer by water displacement (Fa. Ritter or Braud, Bochum-Langendreer, Germany). The biogas amount and composition was analysed with a PACKARD model 427 gas chromatograph equipped with a Micro-WLD-detector and a 1/8 inch Teflon column (1.5 mm i.d. x 1.8 m length) packed with Poropack N (80-100 mesh; Sigma, Deisenhofen). The temperature settings used were as follows: column at 30°C, injector and detector at 250°C. Nitrogen served as the carrier gas at a flow rate of 25 $ml \cdot min^{-1}$.

One hundred μl gas samples were withdrawn from the headspace of the test bottles (in case of reactor, from gas sampling ports) using a Pressure Lok® syringe (Precision sampling Corp., Baton Rouge, Louisiana) and injected on-column into the gas chromatograph. Standard gas (methane, hydrogen and carbon dioxide) was injected under the same conditions to determine the concentration in the samples.

2.1.4 Ammonia

Ammonia was determined by using Method 4500-NH_3 B of APHA (1989) with preceding distillation. The analytical steps were described in Standard Methods (APHA, 1989). The distillation procedure was used to separate the ammonia from interfering substances. Ammonia in the sample was distilled into a solution of boric acid and determined titrimetrically with standard H_2SO_4 with mixed indicator.

2.1.5 Organically Bound Nitrogen (N_{org})

Organically bound nitrogen (N_{org}) was determined by using Method 4500- N_{org} Macro-Kjedahl Method of Standard Methods (APHA; 1989) with the residue after distillation of NH_4-N. Most of organic compounds containing nitrogen were derivatives of ammonia and destruction of the organic portion of the molecule released the nitrogen as ammonia. The Kjedahl method employed sulphuric acid as the oxidizing agent. A catalyst was needed to hasten the oxidation of some the more resistant organic substances. The oxidation proceeded rapidly at temperatures slightly above the boiling point of sulphuric acid (340°C). The boiling point of the acid was increased by addition of sodium or potassium sulphate. When the organic nitrogen has been released as ammonia nitrogen, it was measured in a manner similar to ammonia nitrogen determination as mentioned previously.

2.1.6 pH Value

The pH value of samples was determined potentiometrically with an Ingold pH electrode.

2.1.7 Glucose

Glucose concentrations were determined by using Dinitrosalicylic acid-reagent (Miller, 1959). This method can avoid the dilution error in high concentrations of glucose in the samples (>1000 mg·l^{-1}). A mixture of 0.25 ml sample, 0.25 H_2O bidest. and 1.5 ml reagent was boiled in a water bath (100°C) for 5 min. After cooling, 3 ml H_2O bidest. was added and the colour intensity of the solution was measured spectrophotometrically at 550 nm. A standard curve was prepared for 0 – 700 mg glucose·l^{-1}.

2.1.8 Solid Content (Total Solid-TS and Volatile Solid-VS)

The solid content was determined by using Method 2540 Solids – Standard Methods, (APHA, 1989). The well mixed samples were evaporated in a weighed vessel and dried to constant weight. For determining the total solid (TS), the sample was evaporated at 103 to 105°C. The organic fraction of TS (OTS or volatile solids) was obtained by subtraction of the mineral content (residual ash after oxidation at 550 $\pm$ 50°C) from the total solids content.

2.1.9 Microscopical Examination

The microscopically examination was used for counting the bacterial cell number, observing the shape of bacterial cells and flocs, and also distinguishing the presence of methanogenic bacteria. The number and shape of bacterial cells or flocs was counted with a transmission light microscope with light field or phase contrast. The presence of methanogenic bacteria can be observed using fluorescence microscopy with UV light of 420nm. Under UV light (wave length 420 nm), methanogenic bacteria emitted fluorescence-green colour, due to their content of the hydrogen-transferring coenzyme F_{420}. F_{420} is a deazaflavin that is only occurring in methanogenic bacteria and some Streptomyces species.

All microscopic examinations were proceeded with a Zeiss Axioskop Microscope with transmission light and reflected light fluorescence. For fluorescence observation, the microscope used a UV light which was produced by a short bowed-mercury vapour lamp. This microscope was connected to a digital camera, which was linked directly to a PC for capturing the pictures.

2.1.10 Chemicals

All the chemicals were of analytical grade and were purchased from Merk, Fulka and Sigma Company. Millipore water (MilliQ, Germany) was used for preparing reagent solutions.

2.1.11 Gases

All gases that were used in dispensing oxygen and for gas chromatography were obtained from Linde (Hoellkriegelsreut, Germany).

2.2 Basic Calculations

2.2.1 Hydraulic Retention Time (HRT)

The HRT was the average residence time of the wastewater in the reactor (equation 28). It was the ratio of liquid volume (V_{liquid}) in the reactor and the flow rate (F).

Equ. 28: $$HRT = V_{liquid} / F \qquad \left[\frac{m^3 \cdot d}{m^3} = day \right]$$

2.2.2 Organic Loading Rate (OLR)

The OLR was the amount of organic material, related to COD that was added per liter of reactor volume per day (equation 29).

Equ. 29: $$OLR = c_{inf}(COD) \times F / V_{liquid} \quad \left[\frac{kg \cdot m^3}{m^3 \cdot d \cdot m^3}\right] = \left[kg \cdot m^{-3} \cdot d^{-1}\right]$$

2.3 Experiment Design

2.3.1 Continuous System

2.3.1.1 Laboratory-scale Anaerobic Fixed Bed Reactor (AFBR)

A scheme of the fixed bed reactor, which was used in this study is shown in Figure 6. The primary components of the system were the main reactor compartment, the waste water recycle system, the constant temperature jacket, the feed reservoir, and the liquid and gas effluent system. In addition, the pH could be controlled with an automatic titrator.

This reactor was an anaerobic up-flow fixed bed reactor. The reactor was made of glass with a water-jacketed glass volume of 4.5 l (8 cm i.d., 90 cm length). Supporting material for biomass was arranged inside the reactor. The supporting material was Liapor-clay-polyethylene sinter lamellas (Herding Co., Amberg) with 4 mm thickness that was made of clay and polyethylene polymer (Figure 7). In order to increase the surface area, the material was arranged in radial form (Figure 7). The liquid volume of this reactor was 4.5 l, thus it was completely filled.

Effluent for recirculation was withdrawn slightly below the overflow port and recirculated into the bottom part of the reactor. Fresh feed was pumped into the recirculation stream to maintain a fast mixing. Biogas left the reactor through a port at the top and was measured with a wet gas meter. Thermostated water from a water bath was pumped through the water jacket surrounding the reactor to maintain the temperature at 37°C.

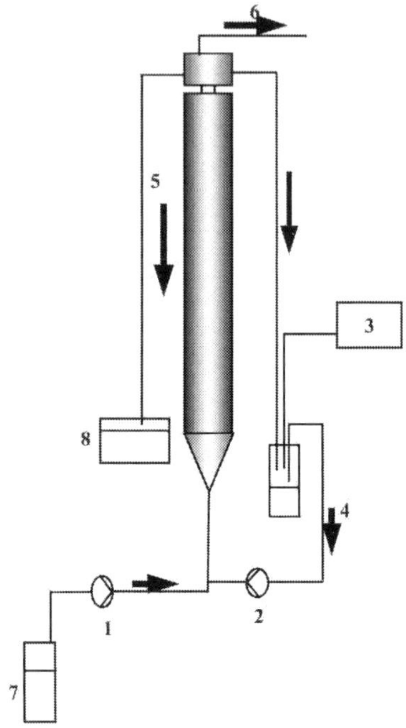

Figure 6 Scheme of anaerobic fixed bed reactor, 1 = peristaltic pump for whey feeding; 2 = peristaltic pump for whey recirculation; 3 = pH-titrator unit; 4 = recirculation; 5 = outlet of whey effluent; 6 = gas line to gas meter; 7 = fresh whey reservoir; → = flow direction, 8 = whey effluent tank.

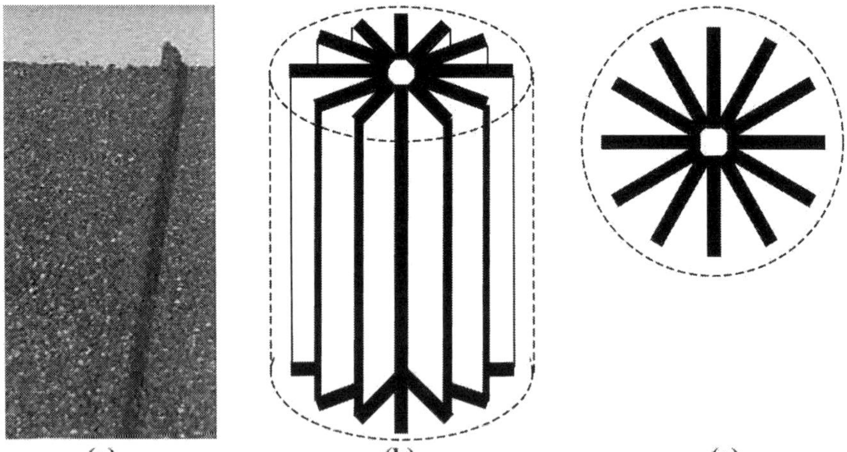

Figure 7 Supporting material inside the AnFB reactor, (a) photograph of a Liapor-clay-polyethylene sinter lamellas (Herding Co., Amberg), (b) side view of the lamellas arrangement, (c) upper view of the lamellas arrangement.

2.3.1.2 Reactor Operation

The reactor was fed with a whey solution, which was prepared from whey powder with a final concentration of 50 g whey·l^{-1}. This reactor was operated under various organic loading rates (OLR) by varying the hydraulic retention time (HRT). The reactor performance was monitored by daily measurement of the pH in the effluent, feed rate and the total gas production. The influent and effluent COD, gas composition and volatile fatty acid contents were measured 2-3 times per week.

The pH in the reactor was controlled automatically with a pH titrator unit, using 1 M NaOH solution, which was added into the recycle line to maintain the pH at 6.8 - 7.0. In certain cases, the pH adjustment system was stopped in order to determine the reactor capabilities for degrading volatile fatty acids.

The reactor performance was improved by adding minerals in this case $FeCl_3$ (technical grade) and vitamin solution (DSMZ-Medium 141).

2.3.2 Batch Experiments

Batch experiments were done in a series of serum bottles with a total volume of 110 ml for each bottle. The inoculums was the effluent of the whey reactor, mixed with additional substrate, to give a total liquid volume of 20 ml, except where stated otherwise. All experiments were done in triplicate. The serum bottles were tightly closed with a rubber stopper and capped with aluminium caps. After displacing the head space air with N_2 or H_2/CO_2, the assay was incubated at 37°C on an orbital shaker.

2.3.2.1 Batch Assay with Different Substrates for Methanogenesis

In these experiments four different substrates were used which were known as precursor substrates for methanogenesis. The substrates that were used included H_2/CO_2

(80%:20% gas mixture), Na-formate, methanol and Na-acetate. These substrates were added in different amounts for determining the kinetics of methanogenesis. The assay variations for each substrate are listed in Table 5.

The reducing agent in this experiment was Na_2S at a concentration of 0.05 $g \cdot l^{-1}$. Resazurin at a concentration of 0.001 $g \cdot l^{-1}$ served as a redox indicator. Trace minerals modified from Medium 141(DSMZ), were used and 10 ml per litre medium were added. The vitamin solution was mentioned in Medium 141(DSMZ and 10 ml per litre medium were added.

Table 5 The conditions of assays for kinetic experiments.

Substrate	Amount in assay	Gas (head space)	Variation
H_2/CO_2	1.00, 1.25, 1.50, 1.75, 2.00 and 2.25 bar	H_2/CO_2	-
Na-Formate (HCOONa)	0, 10, 50 and 100 mM	N_2	-
Methanol (CH_3OH)	0, 10, 50 and 100 mM	N_2	-
Na-Acetate (CH_3COONa)	0, 10, 50 and 100 mM	N_2	- No addition of reducing agent and minerals - Addition of reducing agent (Na_2S and Resazurin) - Addition of reducing agent (Na_2S and Resazurin), trace minerals and vitamins

2.3.2.2 *Batch Assay with Mineral Addition for Degrading the Residual VFAs in the Effluent of the Whey Reactor*

These experiments were carried out using residual VFAs in the effluent of the whey reactor. The concentration of VFAs depended on the condition in the reactor. In these experiments single minerals or combinations of minerals were added in order to improve methanogenesis and acetate degradation. The details of experimental conditions were listed in Table 6.

The reducing agent in this experiment was Na_2S at a concentration of 0.05 $g \cdot l^{-1}$. The redox indicator Resazurin was added to a final concentration of 0.001 $g \cdot l^{-1}$. The trace minerals were those of Medium 141(DSMZ) and 10 ml per litre medium were added. The vitamin solution was that of Medium 141(DSMZ) and 10 ml per litre medium were added.

Table 6 The conditions of assays for volatile fatty acid degradation in effluent of the whey reactor.

Substrate	Amount in assay	Gas (head space)	Variation
Residual VFAs in Whey effluent	15-40 mM	N_2	- Single ions and combinations of ions: final concentration of single ions in the assay = 50 $mg \cdot l^{-1}$ o Fe^{3+} o Ni^{2+} o Co^{2+} o Ni^{2+} and Co^{2+} o Ni^{2+}, Co^{2+} and Mn^{2+} o Ni^{2+}, Co^{2+}, Mn^{2+} and Mg^{2+} o $SO_4^{=}$ - Only Fe^{3+} (50 $mg \cdot l^{-1}$), + reducing agent (Na_2S and Resazurin)

Chapter 3

RESULTS

3.1 Performance of an Anaerobic Fixed Bed (AnFB) Reactor for Treating Whey-containing Wastewater

3.1.1 Organic Removal

The AnFB reactor was fed with an artificial whey containing wastewater with a final concentration of 50 g whey·l^{-1}. This concentration was equal to 50 - 60 g COD·l^{-1}. At a constant concentration of organic feeding, the reactor was run at various hydraulic retention times (HRT) from 2 to 12 days (Figure 8), corresponding to organic loading rates (OLR) of 5-30 g COD·l^{-1}·d^{-1} (Figure 9).

A COD removal efficiency of ≥ 85% was achieved when the reactor was operated at a HRT longer than 5 days and an OLR lower than 10 g COD·l^{-1}·d^{-1}. (Figure 8 and 9). The COD removal efficiency decreased to less than 80% when the OLR was increased above 15 g COD·l^{-1}·d^{-1}.

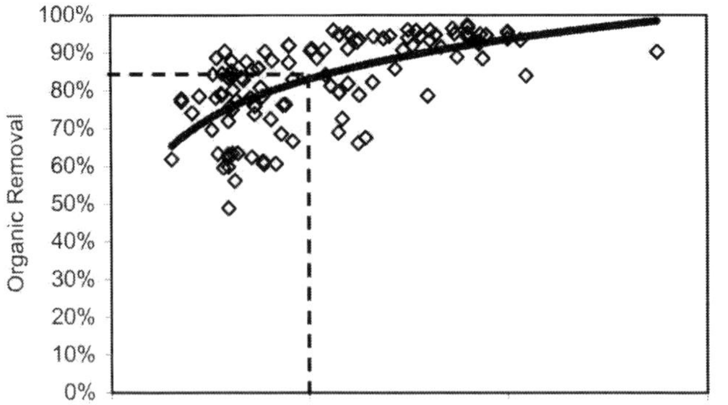

Figure 8 Organic removal efficiency at different HRT.

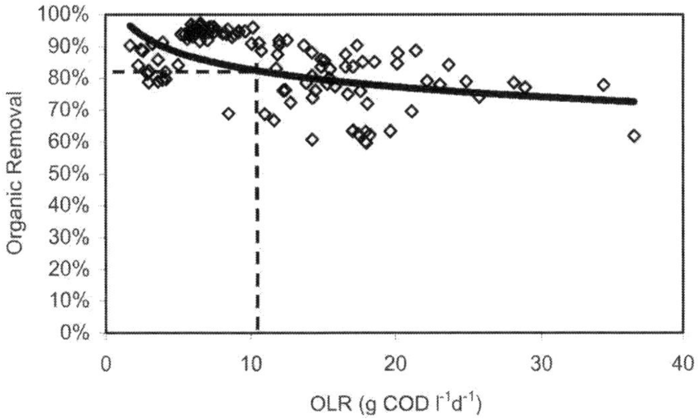

Figure 9 Organic removal efficiency at different OLR.

The results showed that the effluent contained some volatile fatty acids, such as acetate and propionate. These two acids contributed significantly to the residual COD value of the effluent. From the oxidation equation of these organic acids, the COD value of each acid can be estimated. (Equation 30 and 31)

Equ. 30: $CH_3COOH + 2O_2 \rightarrow 2CO_2 + 2H_2O$

Equ. 31: $CH_3CH_2COOH + 3.5O_2 \rightarrow 3CO_2 + 3H_2O$

Based on the above equation 30 and 31, 1 mol·l^{-1} acetic acid (= 60 g·l^{-1}) equals 64 g COD·l^{-1} and 1 mol·l^{-1} propionate (= 74 g·l^{-1}) equals 112 g COD·l^{-1}. Using these constants, the composition of organic compounds in the effluent of the whey reactor was specified for the contribution of single compound (Figure 10).

A higher OLR tended to increase the residual organic compounds in the effluent. At a lower OLR (less than 10 g COD·l^{-1}·d^{-1}), the effluent contained only acetate and other non-acid organic compounds, while acetate and propionate appeared as components of the organic material in the effluent of the whey reactor at a higher OLR. Both acetate and propionate seemed to increase with increasing OLR (Figure 10).

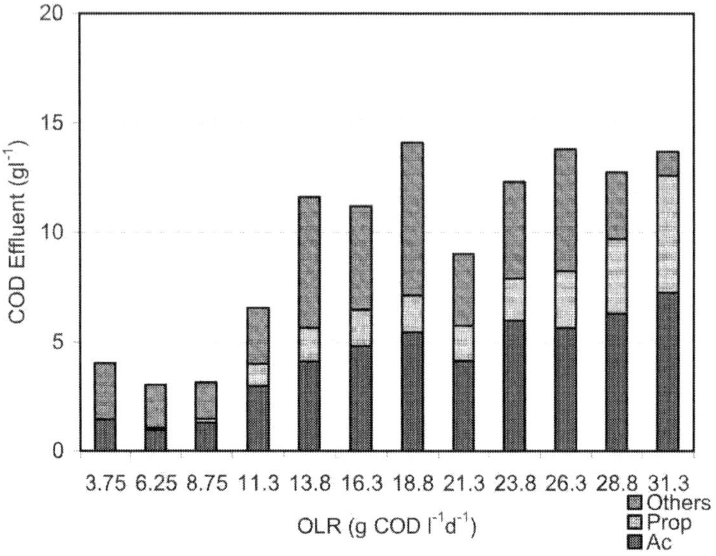

Figure 10 Composition of COD in the effluent.

3.1.2 Residual Volatile Fatty Acids

Volatile fatty acids are the end products of hydrolysis and acidogenesis in the multistep nature of anaerobic digestion. These products are utilized by acetogenic and methanogenic bacteria to produce methane. When the rates of acetogenesis and methanogenesis are compared with the rate of acidogenesis, the growth rate of methanogenic and acetogenic bacteria was always slower than that of acidogenic (fermentative) bacteria. The rates of acetogenesis and methanogenesis determine the whole anaerobic digestion process rate, which means that the HRT has to be adjusted such that both phases, in particular the methanogenic phase is not overloaded.

The residual non-degraded fatty acids in the effluent of the methane reactor contribute significantly to the remaining COD in the reactor effluent. Figure 11 shows the relationship between concentration of fatty acids and remaining COD in the effluent. Higher concentrations of acetate and propionate in the effluent corresponded with higher COD values in the effluent.

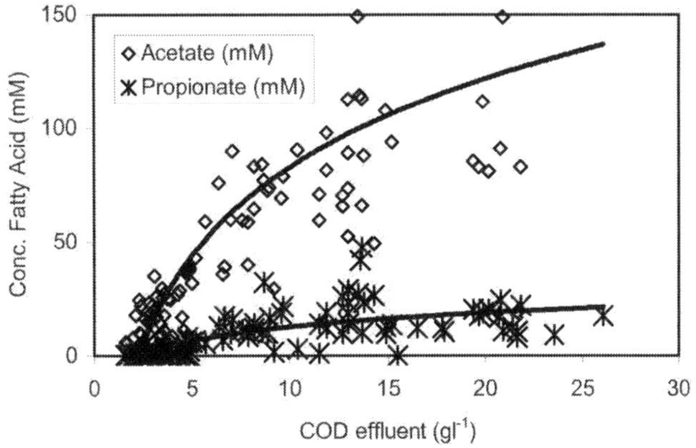

Figure 11 Residual volatile fatty acids concentrations at different COD in the effluent of the whey reactor.

The concentration of residual volatile fatty acids in the effluent of the whey reactor, such as acetate, propionate, butyrate and valerate, indicated a reduction of the removal efficiency. This was due to the imbalance between acid formers and acetogenic and methanogenic bacteria in the system. This condition can trigger a further disturbance in reactor operation. A rapid acid production, which is not followed by methane production may lead to accumulation of fatty acids and lower the pH inside the reactor. This unfavourable condition can inhibit the growth of methanogenic bacteria and lead to failure of the biogas reactor.

The concentration of fatty acids in the effluent during the experiments can be seen in Figure 12 and 13. A complete conversion of fatty acids to methane requires a longer HRT than 10 days. At a HRT less than 6 days, the concentration of acetate in the effluent was higher than 50 mM. At a higher OLR than 10 g $COD \cdot l^{-1} \cdot d^{-1}$, the acetate concentration tended to increase to more than 50 mM (Figure 13). The optimal operating condition for the whey reactor with a COD concentration of 50 $g \cdot l^{-1}$ was a HRT between 4 to 8 days and an OLR of less than 10 g $COD \cdot l^{-1} \cdot d^{-1}$.

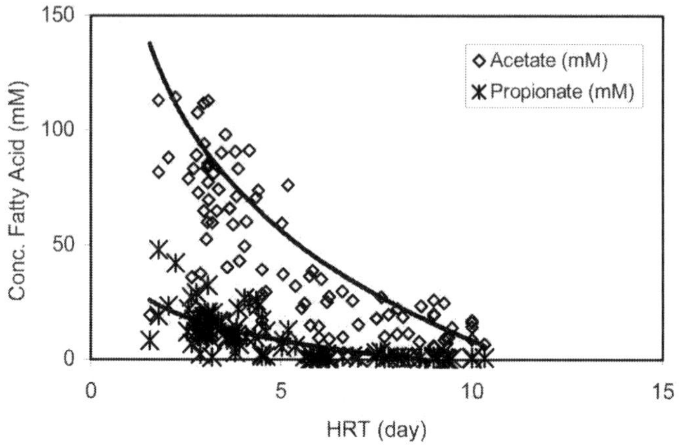

Figure 12 Residual volatile fatty acids concentration at different HRT.

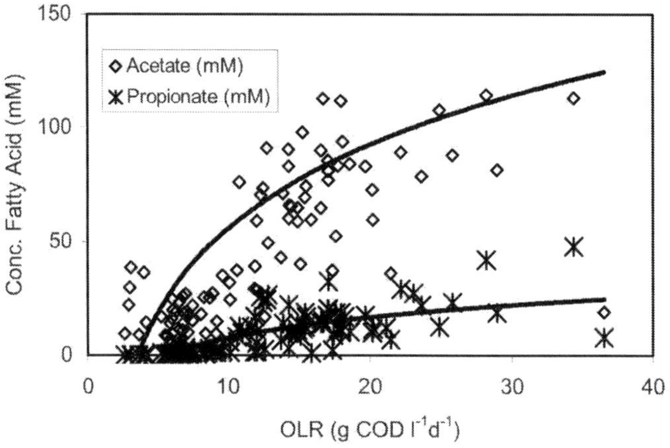

Figure 13 Residual volatile fatty acids concentration at different OLR.

3.1.3 Methane Production

During the anaerobic degradation of organic material, methane is produced as one of the major end products. The other major product is CO_2. Anaerobic conversion of sugars like glucose can theoretically yield about 0.35 L_{STP} of methane per 1 g of COD, as derived from the Buswell equation (Equ. 32) or the conversion stoichiometry.

Equ. 32: $$C_nH_aO_b + (n - \frac{a}{4} - \frac{b}{2})\ H_2O \rightarrow (\frac{n}{2} - \frac{a}{8} + \frac{b}{4})\ CO_2 + (\frac{n}{2} + \frac{a}{8} - \frac{b}{4})\ CH_4$$

The change of methane production rates during the experiment for decreasing HRT is presented in Figure 14 and for increasing OLR in Figure 15. The methane production rate decreased at longer HRT which corresponded with lower OLR. The optimal methane production rate was reached at a HRT of around 5 days with an OLR of 10 g $COD \cdot l^{-1} \cdot d^{-1}$.

The specific methane production is the amount of methane that is produced from each gram COD which is added to the system. The results of these experiments were plotted in Figure 16. The specific methane production values varied with the OLR. A higher OLR tended to lower the specific methane production.

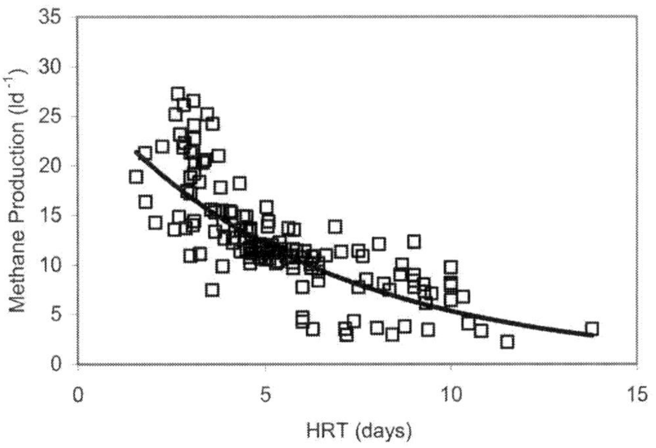

Figure 14 Methane production at different HRT.

At different OLR, the specific methane production was in the range of 0.15 to 0.35 l CH_4 per g COD removed. The specific methane production, which was close to the theoretical yield of methane per 1 g of COD that was degraded was reached at an OLR lower than 10 g $COD \cdot l^{-1} \cdot d^{-1}$. Increasing the organic loading up to 35 g $COD \cdot l^{-1} \cdot d^{-1}$ reduced the specific methane production to les then 40% of the theoretically expected amount.

The decreasing of specific methane production had the same tendency as the results of COD removal (Figure 16). When the intermediately excreted fatty acids were converted to methane, the concentration of residual organic compounds in the effluent was low and the percentage of COD removal was high.

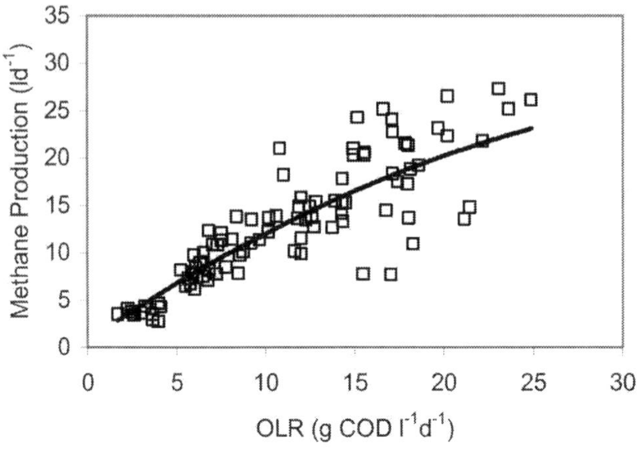

Figure 15 Methane production at different OLR.

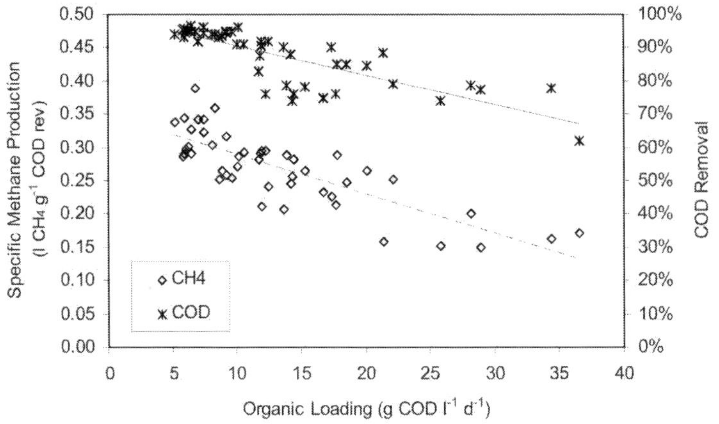

Figure 16 Specific methane production and COD removal at different OLR.

With the knowledge of the total substrate concentration and the carbon content of each substrate, a carbon mass balance of the methanogenic step can be obtained. The dominant reactions in the methanogenic system are: (equation 33 – 36)

Equ. 33: $CH_3CH_2COOH + 2\ H_2O \rightarrow CH_3COOH + CO_2 + 3H_2$

Equ. 34: $CH_3(CH_2)_2COOH + 2\ H_2O \rightarrow 2CH_3COOH + 2\ H_2$

Equ. 35: $CH_3COOH \rightarrow CH_4 + CO_2$

Equ. 36: $4H_2 + CO_2 \rightarrow CH_4 + 2\ H_2O$

Under the (reasonable) assumption that no enrichment of H_2 occurred in the system, as long as no inhibition takes place, the following two overall reactions (equation 37 and 38) for the degradation of propionic and butyric acid result. In other words, the H_2 formed from VFA in reaction 33 or in reaction 34, respectively, is completely consumed in reaction 36.

Equ. 37: $4CH_3CH_2COOH \rightarrow 7CH_4 + 5CO_2 + 2H_2O$

Equ.38: $2CH_3(CH_2)_2COOH \rightarrow 5CH_4 + 3CO_2$

According to the above equations, the amount of methane from the residual volatile fatty acids in the effluent can be calculated. The conversion constants are 1 mole CH_4 per mole acetate and 1.75 mole CH_4 per mol propionate.

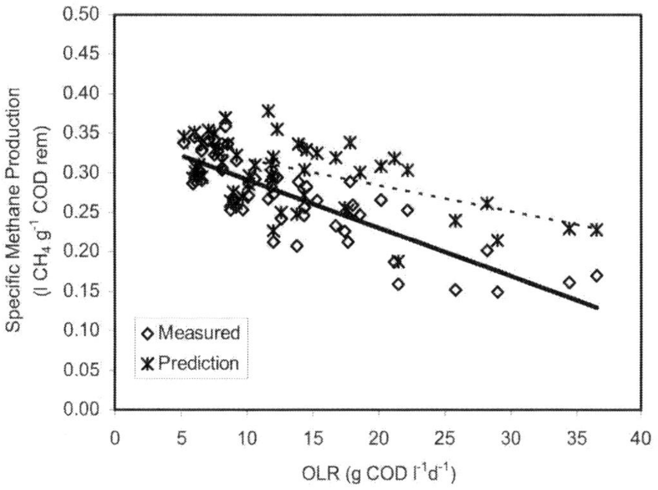

Figure 17 Specific methane production and total predicted methane production.

When all volatile fatty acids in the effluent were converted to methane, the prediction of the total specific methane production is between 70% to 100% from the theoretical amount.

3.2 Performance of an Anaerobic Fixed Bed (AnFB) Reactor for Treating Whey-containing Wastewater at a Controlled pH

3.2.1 Effect of HRT on pH and Concentration of VFAs in the Reactor

In order to find out the capabilities of the AnFB reactor for accommodating the change of pH in the reactor, the AnFB reactor was run at different HRT (4, 8 and 12 days) without the pH-regulator and fed with whey solution at a concentration of 50 g whey·l^{-1}. In the absence of pH-adjustment, the pH in the reactor fluctuated due to the production and utilization of volatile fatty acids. The results of these experiments are presented in Figure 18.

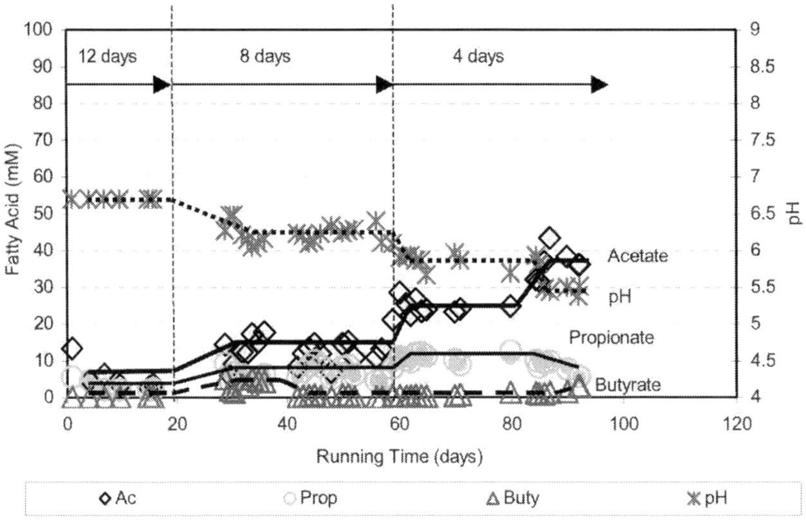

Figure 18 Effect of HRT on pH and production of VFAs in AnFB reactor.

In Figure 18, the pH in AnFB reactor decreased with shorter HRT. The average pH at a HRT of 4, 8 and 12 days was 5.7; 6.25 and 6.7 respectively. At the same time, the concentration of volatile fatty acids, in this case acetate, propionate, and n-butyrate were increasing with the shorter HRT. The concentration of acetate went up from 5 mM at a HRT of 12 days to 38 mM at a HRT of 4 days, while for propionate the concentration increased slightly from 3 mM at a HRT of 12 days to 10 mM at a HRT of 4 days. The concentration of n-butyrate did not have a certain trend with a change of HRT. The concentration of volatile fatty acids escalated when the acid was no longer degraded and accumulated in the reactor. This situation happened when the acid production rate was higher than the acid utilization rate. The imbalanced production and utilization rates showed that the population of acetogenic and methanogenic bacteria was effected by the change of the environment due to shortening the HRT. This result agreed with Ueno et al. (2001) who stated that the intermediate metabolites of methane fermentation such as lactate, ethanol, acetate, butyrate, formate, hydrogen, and carbon dioxide, were accumulated as dilution rate increase.

At a shorter HRT, the methanogenic and acetogenic bacteria, which grow slowly, were incapable to degrade the products of acidogenesis. The unconverted acids remained in the system. When the acidogenesis still proceeded, the acids accumulated in the system. An increase of the concentration of acids decreased the pH in the reactor. A pH lower than 6.8 was unsuitable for methanogenic bacteria. This situation disturbed the performance of methanogenesis from whey-containing wastewater completely.

Acetate was the dominant accumulating fatty acid compared to other volatile acids. It pointed out that the conversion rate of acetate to methane was interfered by either the number of acetogenic and methanogenic bacteria, which were not sufficient, or the environmental conditions were not favourable for these species.

The population of microorganisms in the reactor seemed to be capable for degrading the accumulated propionate in the reactor. It happened after adaptation at a HRT of 4 days that the concentration of propionate decreased from 10 mM to 5 mM. At the same time the concentration of acetate increased. This showed that the microorganisms converted propionate to acetate (Figure 18, from day 80 on).

In general, the concentrations of fatty acids were increasing with decreasing HRT. Usually right after the HRT was reduced, the concentration of fatty acids increased but after several hours or days the concentration became constant. This could be the adaptation period for acid converting bacteria which was the longer the shorter the HRT was adjusted. When the acid concentration was constant, the population was capable to establish a new equilibrium.

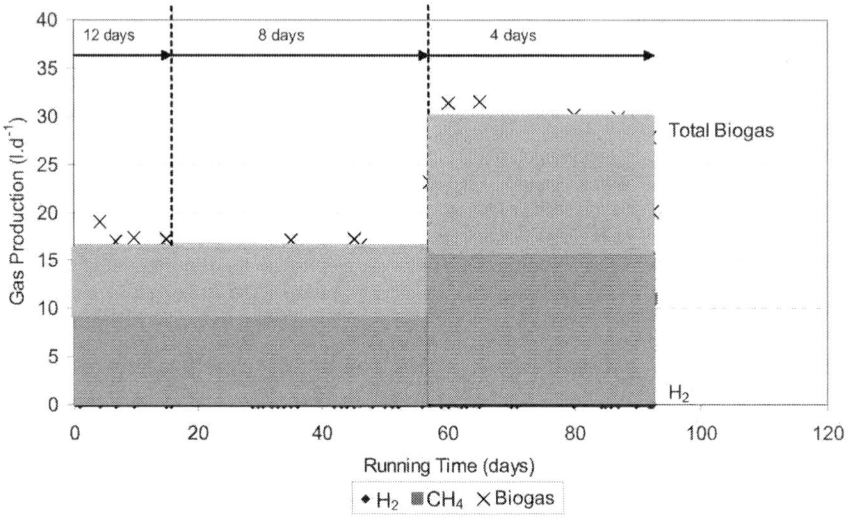

Figure 19 Effect of HRT on biogas production rate in AnFB reactor.

Varying the HRT did not only influence the pH and concentration of fatty acids in the effluent, but also the gas production. The results pointed out that the alteration of the HRT from 12 days to 8 days did not give a significant change in gas production (methane and total biogas). If the HRT was reduced from 8 days to 4 days, the biogas production almost doubled for both, methane and total biogas. These results were synchronous to previous result which stated that higher HRT tended to decrease the methane and biogas production. It is still unclear why there was no considerable difference in gas production at a HRT of 12 days and of 8 days. This had mentioned also by Ueno et. al. (2001) that stable methane fermentation, producing methane and carbon dioxide was observed at relatively low dilution rates (less than 0.03 d^{-1} on glucose).

Anaerobic conversion of propionate and butyrate must produce hydrogen gas. Since the hydrogen was not detected in the biogas, there was no enrichment of hydrogen in the reactor. All the hydrogen from acid conversion was used directly by hydrogen-utilizing methanogenic bacteria to produce methane. This condition was suitable for further conversion of higher concentration of propionate and butyrate to acetate, CO_2 and H_2, because enrichment of hydrogen in the system, which could inhibit the conversion of these two acids, was not accumulated.

3.2.2 Effect of Feeding on Degradation of Accumulated VFAs in an AnFB Reactor

While running the reactor with a HRT of 4 days, the effect of feeding on the performance of the reactor was observed. Stopping the feeding meant that there was no substrate for acid production in the reactor. In this case, the decrease of the acetic, propionic and butyric acid concentration represented the capabilities of acetogenic and methanogenic bacteria in utilizing fatty acids.

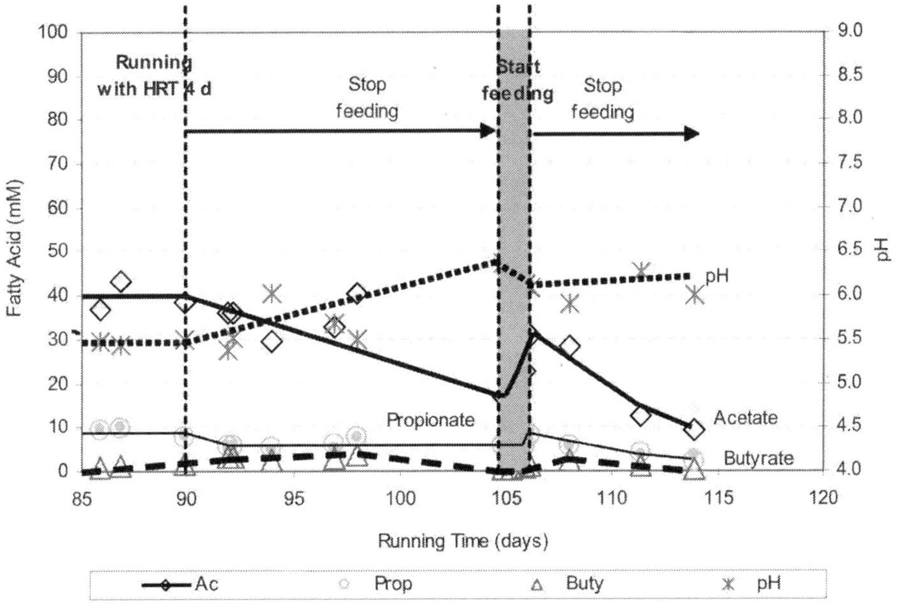

Figure 20 Effect of changes of feeding on degradation of accumulated VFAs in AnFB reactor.

At the first stop of feeding, once the feeding was terminated, the reactor became a batch system since there was no input and output. Within this period, the pH inside the reactor went up from 5.5 to 6.4 (Figure 20). At the same time, the concentration of acetate and the propionate decreased but butyrate was slightly increasing. The acetate was successfully degraded by 22 mM in 15 days, propionate by 5 mM in 15 days and butyrate by 3 mM in 7 days.

When the system revealed a capacity of acid degradation, feeding of the reactor was started again with a concentration of 50 g whey per litre at a HRT of 4 days. As soon as the fresh influent entered the system, the pH inside the reactor started to decline. This was a sign of accumulation of fatty acids in the reactor. Within 2 days the concentration of acids increased by 12 mM for acetate, 5 mM for propionate and 2 mM for butyrate. The feeding of 50 g whey·l^{-1} at a HRT of 4 days (= 12.5 kg·m^{-3}·d^{-1}) was beyond the capabilities of the population in the reactor for consuming the products of the acidogenic stage.

The results from the first and second interruption of feeding showed, that the reactor had a potential for degradation of fatty acid from previous feeding. Its acid degradation rate was less than the acid production rate. This imbalanced condition led to a build up of fatty acid concentrations in the reactor. In order to ascertain this hypothesis, the third experiment was preceded by terminating the feeding. Once the fresh input was stopped, the acid concentration went down with faster rate than in first phase. In 7 days, the acetate decreased by 20 mM, propionate by 7 mM and butyrate by 2 mM. At the same time the pH raised again. This result indicated that the acid converting bacteria, in this case acetogenic and methanogenic bacteria were available in the system and their conversion rates were less than the acid production rates in the reactor. The complete results of process rates in these experiments can be seen in Table 7

Table 7 The reduction and accumulation rate of VFAs in the reactor.

Volatile Fatty Acids	**Unit**	**First Stop**	**Feeding**	**Second Stop**
		0 g whey·l^{-1}	**50 g whey·l^{-1}**	**0 g whey·l^{-1}**
		I	**HRT 4 days**	II
Acetate	mmol·h^{-1}	- 0.275	1.125	- 0.536
Propionate	mmol·h^{-1}	- 0.063	0.469	- 0.188
Butyrate	mmol·h^{-1}	- 0.038	0.188	- 0.054

From the Table 7, the accumulation rates for all fatty acids during feeding are higher than their reduction rates during starvation period. This indicates that the acetogenic and methanogenic bacteria metabolize and grow slower than the acidogenic bacteria. The acid degradation rates in the first starvation phase were lower than that in the second starvation phase. If a third feeding/starving cycle was induced after day-114 (not shown) the fatty acid production rates were the same as those in Table 7, while the fatty acid degradation rates during starvation were again improved. It seemed that the degradation rates were influenced by environmental conditions and the metabolic capabilities of the acetogenic/methanogenic population still could be improved, as had also been mentioned by Sánchez et.al. (2000) that the apparent kinetic constants of the biomethanization process increased 2.3 times when the intial pH of the influent was increased from 7.0 to 7.6. The pH and concentration of acids in the third phase were suitable for acetogenic and methanogenic bacteria which improved their capabilities for degrading the acids.

3.2.3 Effect of Feed Interruption on Reducing the Accumulation of VFAs in an AnFB Reactor

The previous experiments showed that HRT and feeding concentration contributed to the accumulation of fatty acids in the whey reactor. In this experiment, the feeding was stopped during steady state at 4 d HRT and the AnFB operated as a batch reactor. During the observation period, the biogas production, pH and concentration of fatty acids were monitored. The results are presented in Figure 21.

At the beginning the reactor was in an acidic condition (or running at its limits) and accumulated acetate, propionate and butyrate. At this point the pH was 5.5, the fatty acids were 47 mM for acetate, 15 mM for propionate and 5 mM for butyrate, and biogas production was still 27 $l \cdot d^{-1}$ (more than 5 $l \cdot l^{-1} \cdot d^{-1}$). After the feeding was stopped, the conditions in the reactor were monitored within the next 6 days. Right after no fresh influent entered into the reactor, the pH started to increase. It went up from 5.5 to 7.25 and remained constant at around 7.25 within one and a half day. This meant that the microorganisms activities inside the reactor were adapted to high-rate fatty acid degradation e.g. acetate, that would be generated from butyrate or propionate degradation. There was no further accumulation of fatty acids in the reactor. This argument is proved by the course of fatty acid concentrations. Within 6 days, no residual fatty acids were found in the effluent if feeding was started again. Acetate and butyrate were decreased around 50% after one day running without feeding, while propionate was only decreasing less than 30%. Acetate and propionate were completely used up in 6 days, while butyrate was used up in 2 days. The maximal reduction rates for all fatty acids are listed in Table 8. Comparing with the results, the acetate and butyrate reduction rates were faster, and the propionate reduction rate was slower. The different condition between this experiment and the previous experiment was the pH inside the reactor. In the previous experiment, the maximum pH after recovery was 6.4 while in this experiment, it was 7.25. The pH of the environment seemed to have a significant effect on the degradation rate of fatty acids. Degradation of acetate and butyrate went faster at a pH around neutral (= pH around 7.0) while propionate degradation was faster in relatively acid pH (= pH around 6.5).

No fresh feeding did not only have an effect on degradation of accumulated of fatty acids, but it also interfered with the biogas production. The experiment showed that the biogas production was decreasing with increasing running time and finally it stopped on the forth day. The biogas is a product of the activities of methanogenic bacteria. These bacteria used short fatty acids such as formate or acetate and H_2/CO_2 as their substrates for producing biogas. With no new substrate supplied for acidogenic and acetogenic bacteria into the reactor, acetate and H_2/CO_2 were finally running out for biogas production. A limited amount of fatty acids reduced the biogas production to a rate of only 0.05 $l \cdot h^{-1}$. Residual concentrations of fatty acids of less than 3 mM apparently did not have significance for biogas production. These were the threshold concentration of acetate utilization by *Methanosarcina sp*. while *Methanosaeta sp* would leave behind only ≥ 0.5 mM acetate (Stams et.al., 2003). Consequently, the whey reactor must contain *Methanosarcina sp*. In the biofilm on the Liapor-polyethylene sinter lamellas for methanogenesis from acetate.

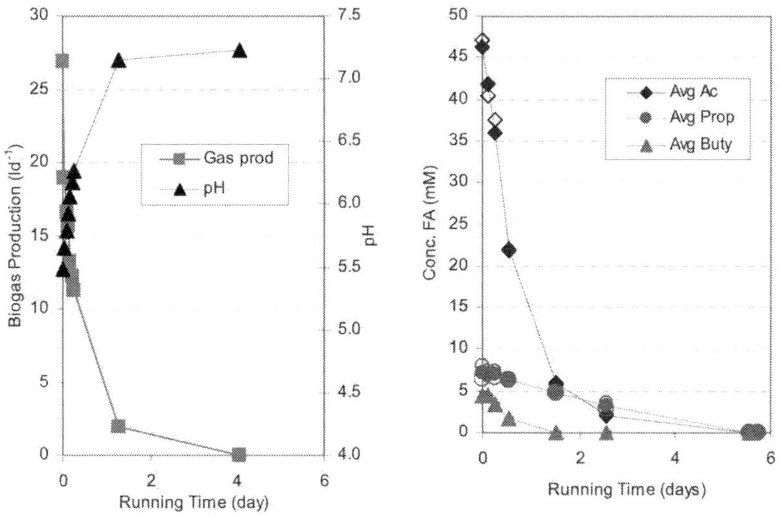

Figure 21 Effect of no feeding on pH, production of biogas and degradation of VFAs in an AnFB reactor.

Table 8 The reduction rate of VFAs in the reactor.

Parameter	Unit	Rate
Acetate	$mmol \cdot h^{-1}$	- 1.05
Propionate	$mmol \cdot h^{-1}$	- 0.05
Butyrate	$mmol \cdot h^{-1}$	- 0.17
Biogas Production	$l \cdot h^{-1}$	- 0.05
pH	$unit \cdot h^{-1}$	+ 0.05

3.2.4 Effect of a HRT of 4 and 6 days on the Accumulation of VFAs in an AnFB Reactor after Fatty Acid Degradation by Interruption of the Feeding

The population in the whey-degrading reactor had shown their capabilities in reducing the accumulation of fatty acids. However, the experiment in Section 3.2.2 revealed that the population was not able to prevent the occurrence of fatty acids accumulation when the reactor was fed continuously with fresh substrate. Two experiments were set up to find out the maximal activities of the population, when new substrate continuously was supplied: one with a HRT of 4 days and the other with a HRT of 6 days. The results are presented in Figure 22 for a HRT of 4 days and in Figure 23 for a HRT of 6 days. Both experiments were run after the accumulation of fatty acids in the reactor has been eliminated by changing the system for a short period into a batch system. The continuous system was run and the accumulation of fatty acids was monitored along the running time. The monitoring period was 6 days.

The first run was with a HRT of 4 days. The whey inside the reactor did not contain significant amount of accumulated fatty acids. The final concentrations of fatty acids were 7 mM for acetate, 4 mM for propionate and 0 mM for butyrate. The pH inside the reactor was 7.0 and the biogas production was 5 $l \cdot d^{-1}$. The feeding was whey solution at a concentration of 50 g whey$\cdot l^{-1}$.

After the feeding was supplied to the reactor, biogas was produced increasingly. The rate of gas production was 0.03 $l \cdot h^{-1}$. These positive results were followed by a decrease of the pH value. It went down from 7.0 to 6 in one day with the rate around 0.04 units per hour. The condition inside the reactor became acidic afterwards and the pH was further decreasing to 5.5 at the sixth day. The acidification condition was due to a higher production of fatty acids than degradation during the process. In a period of 6 days, the concentrations of fatty acids were increasing. The maximum concentrations of the fatty acids were 41 mM for acetate, 8 mM of propionate and 4 mM of butyrate. At this moment, the continuous running was stopped to prevent further disturbance of the population of microorganisms inside the reactor.

During these experiments, acetate was the predominant fatty acid that accumulated in the reactor. Although its concentration was still under the toxic concentration for methanogenic bacteria, the increase of the acetate concentration of around 22.5 mM reduced the pH value around 1 unit. This condition was unfavourable for the methanogenic bacteria. In such situation, the activities of methanogenic bacteria could be inhibited or at least slowed down which would interfere with the conversion of fatty acids to biogas. When this happened, the accumulation of fatty acids raised. The apparent production rates of fatty acids during the experiment were listed in Table 9.

It has, however, to be considered that these are the accumulation rates, resulting from acetate production from lactose and acetate degradation by presumably *Methanosarcina sp*. If one would take the acetate degradation rate of Table 7 an acetate production rate of 0.646 would result (degradation rate + accumulation rate = production rate).

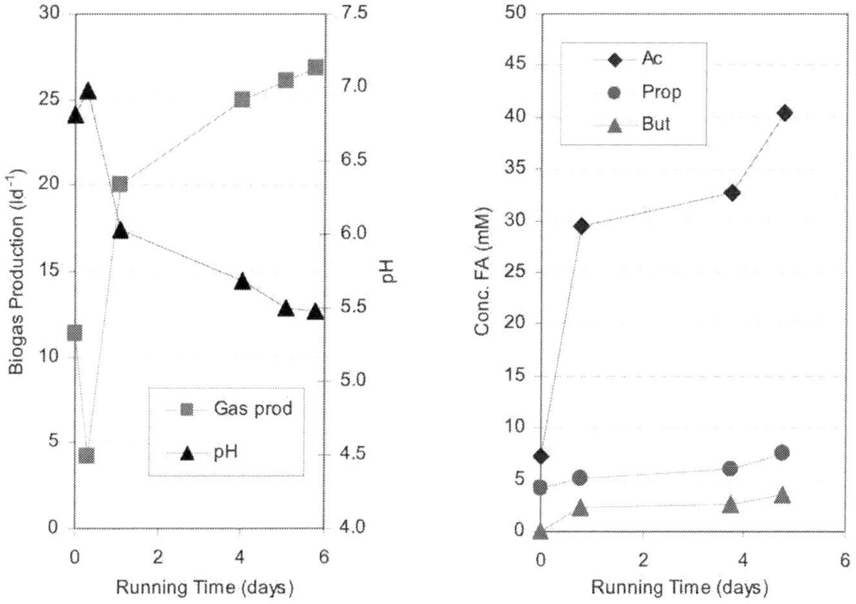

Figure 22 Effect of a HRT of 4 days on pH, production of biogas and accumulation of VFAs in an AnFB Reactor.

Table 9 The accumulation rates of VFAs in the reactor with a HRT of 4 days after interruption of feeding.

Parameter	Unit	Rate
Acetate	$mmol \cdot h^{-1}$	+ 0.11
Propionate	$mmol \cdot h^{-1}$	+ 0.03
Butyrate	$mmol \cdot h^{-1}$	+ 0.03
Biogas Production	$l \cdot h^{-1}$	+ 0.03
pH	$unit \cdot h^{-1}$	- 0.04

The second run was with a HRT of 6 days. The reactor was fed with the same concentration of whey solution (50 g whey$\cdot l^{-1}$). The initial pH was 7.25, biogas production was 0 $l \cdot d^{-1}$. The concentration of fatty acids was close to zero for every acid. Right after the new feeding was supplied to the reactor, the production of biogas started. The biogas production rate was 0.02 $l \cdot h^{-1}$. At the same time the pH inside the reactor decreased. It dropped by 1 unit to 6.25 within one day. In this period, the accumulation of fatty acids had occurred. The acetate concentration was increased to for about 25 mM. It was almost the same as during a HRT of 4 days. For an increase of the acetate concentration by 1 mM in both experiments, the pH inside the reactor dropped around 0.04 units.

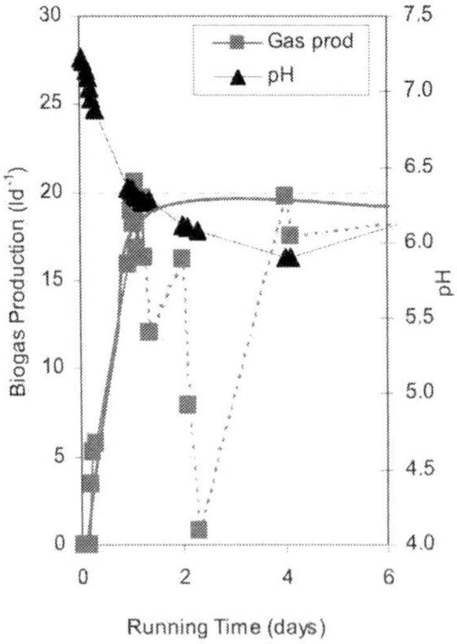

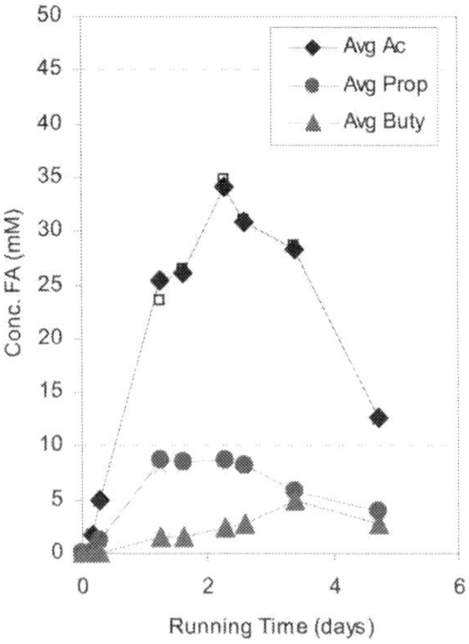

Figure 23 Effect of a HRT of 6 days on pH, production of biogas and accumulation of VFAs in an AnFB reactor.

Acetate was accumulating rapidly in the reactor. It had reached 35 mM within 2 days. Acetate accumulation was followed by propionate accumulation that reached 8 mM in one day. Butyrate was accumulating more slowly compared the other two acids. It reached 5 mM in four days. The accumulation rates of the fatty acids were listed in Table 10.

Table 10 The accumulation rates of VFAs in the reactor at a HRT of 6 days.

Parameter	**Unit**	**Rate**
Acetate	$mmol \cdot h^{-1}$	+ 0.58
Propionate	$mmol \cdot h^{-1}$	+ 0.25
Butyrate	$mmol \cdot h^{-1}$	+ 0.06
Biogas Production	$l \cdot h^{-1}$	+ 0.02
pH	$unit \cdot d^{-1}$	- 0.03

Up to the forth day, the pH inside the reactor has been decreasing to 5.9, but after that it went up again. It was due to a degradation of accumulated fatty acids. The concentration of acetate decreased from day 2 on propionate from day 3 and butyrate after day 3. This indicated that at a HRT of 6 days, the conversion rate of fatty acids was equal or even greater than the rate of acid production. This led to a steady state of the system since the accumulation of fatty acids and a drop of the pH could be prevented. The first four days could be the adaptation period for methanogenic bacteria. They were active enough to consume the acetate and H_2/CO_2 which were produced in degradation of whey by acidogenic and acetogenic bacteria.

The conversion of fatty acids to biogas increased gradually as deduced from the amount of biogas production. The biogas production readings were out of range (day 2 and day 3) because there was a trouble with the gas meter. In Figure 23, they were signed with dash mark. The biogas production was estimated to be constant after the first day. This could be expected as the activity of methanogenic bacteria has reached its maximum capacity. The methanogenic bacteria had their full capacity; they converted fatty acids at the same rate at which acid production occured.

From these two experiments, the HRT or the organic loading became a parameter that determined the concentration of fatty acids in the effluent. A shorter HRT led to higher fatty acid production rates, fatty acids accumulation and a drop of the pH which reduced the ability of methanogenic bacteria in converting all fatty acids to biogas.

The last experiments showed that a long enough HRT could prevent the accumulation of fatty acids in the reactor. This argument was also proved by the following results, when there was a disturbance due to an oxygen exposure accident. This happened during repairing the recirculation tube. After the population was contacted with oxygen, the activity of methanogenic bacteria ceased almost completely. The fatty acids concentration, increased and the pH dropped, indicating a higher oxygen tolerance of acidic bacteria.

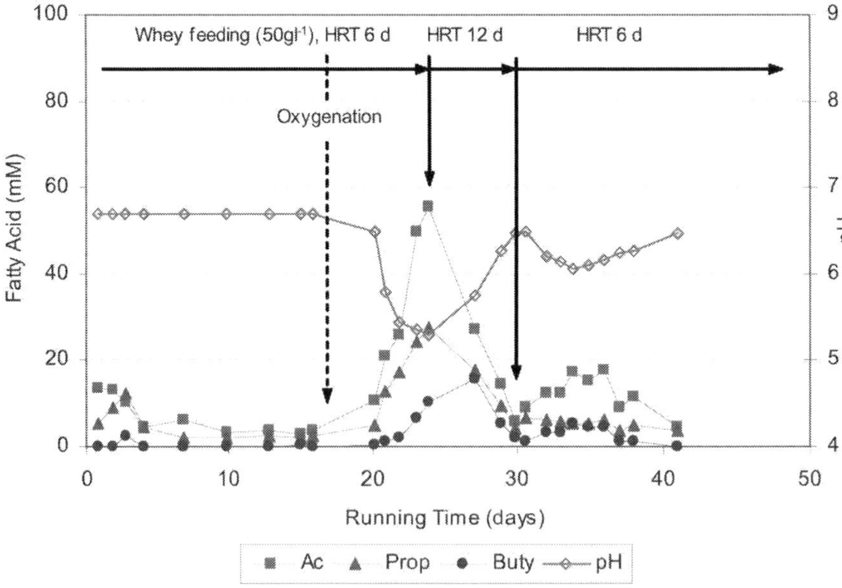

Figure 24 Effect of varying HRT on reducing the accumulation of VFAs in the reactor after oxygenation.

When the reactor was run at a HRT of 6 days with a feeding of 50 g whey$\cdot l^{-1}$, the concentrations of fatty acids were constantly less then 10 mM and the pH inside the reactor was ≥ 6.8. Right after oxygenation happened, the pH decreased from 6.8 to 5.2 within 7 days. At this time, the concentrations of fatty acids were 55 mM for acetate, 25 mM for propionate and 10 for butyrate. This condition threatened the population in the reactor. At these conditions, the HRT of 6 days was reduced to 12 days in order to give the methanogenic bacteria a chance for recovering. The results are presented in Figure 24.

This effort was success as the reactor could establish a new balance of metabolism of the complex population. This was indicated by the increase of the pH inside the reactor. Within 6 days, the pH almost reached its initial value and the concentrations of fatty acids were less than 10 mM. It showed that the methanogenic bacteria need a longer HRT for recovering their activity. The methanogenic bacteria which are known as slower growing microorganism need a longer time to get over the damage by oxygen.

Once the conditions in reactor were back to normal, the HRT was set back to 6 days of HRT. This change intermittently led to a slight accumulation of fatty acids and a decrease of the pH value. However, this situation occurred only for a short time and the concentration of all fatty acids remained below 20 mM. Under these conditions, the methanogenic bacteria were able to maintain there metabolic activity and to survive and they could carry out the conversion of fatty acids at a sufficient rate.

3.2.5 Effect of pH Regulation on the Accumulation of VFAs in an AnFB reactor

The population of the methanogenic bacteria is very sensitive to pH changes. The fluctuation of pH values can influence and even prevent the growth of these bacteria. This group of bacteria prefers to grow at neutral or slightly alkaline pH conditions. A pH regulator system was used for maintaining the pH inside the reactor around neutral. Alkaline solution was titrated automatically into the reactor when the pH of the effluent was less than 6.9. Sodium hydroxide was the caustic solution that was used during the experiment.

The effect of pH regulation on reactor performance was observed at a HRT of 4 days, a HRT at which the population of microorganisms inside the reactor was uncapable to scope with the accumulation of fatty acids. The results of the experiment without automatic pH-titration are presented in Figure 25. At the beginning the reactor was run with a HRT of 4 days with 50 g whey·l^{-1} as the feeding solution and with a pH regulation system. During the initial running, accumulation of fatty acids occurred. The concentration of fatty acids was 140 mM for acetate, 10 mM for propionate and 25 mM for butyrate. At this time the pH was controlled at 7.0 by a pH regulation system. In order to eliminate the previously accumulated fatty acids, the feeding was stopped together with the pH titrator. Within 12 days, the concentrations of fatty acids were reduced to less than 10 mM for acetate and almost none for propionate and butyrate. The pH inside the reactor increased to 8.5.

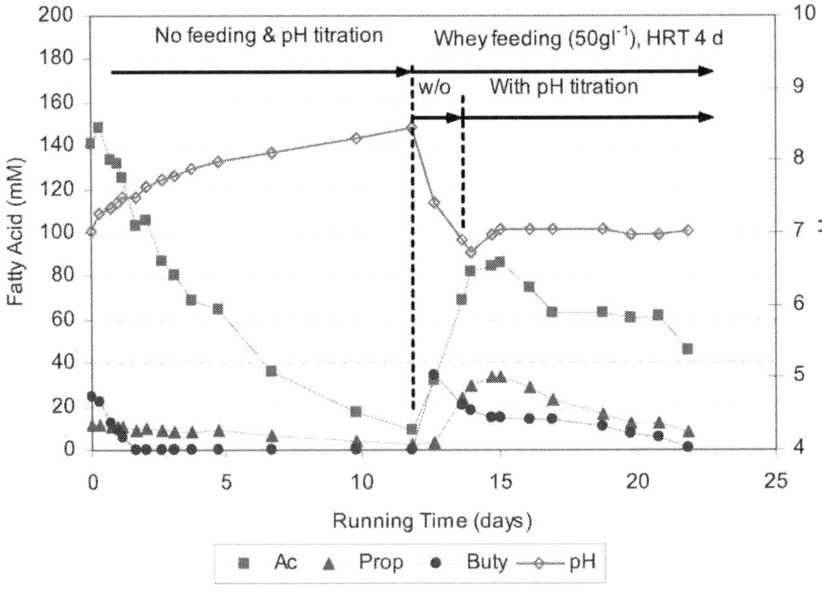

Figure 25 Effect of pH adjustment on reducing the accumulation of VFAs in the reactor.

After the proper condition of methanogenesis was available, the feeding system was started with the same concentration of whey solution at a HRT of 4 days and without pH titrator (Figure 25, day 12). Once the feeding was started, accumulation of fatty acids was detected. Within 3 days, the pH value inside the reactor has dropped by 1.7 units and the concentrations of fatty acids were 80 mM for acetate, 35 mM for propionate and 15 mM for butyrate. At this time the concentration of butyrate began to decrease. This indicated that the butyrate conversion to shorter fatty acids could proceed. At the same time the concentration of propionate and acetate were also increasing. In order to prevent further accumulation of acetate and propionate, the pH regulation system was started at day 15.

When using the pH regulation, the pH inside the reactor was kept around 7.0. As a result of this optimum pH condition for methanogenic bacteria, the concentration of fatty acids started to go down, and within 7 days half of the concentration of fatty acids had been converted to biogas. This pH condition was the optimum for whey treatment at a HRT of 4 days. At steady state conditions the concentration of acetate was around 40 mM, and of around 10 mM for propionate.

This result proved that pH regulation could prevent the accumulation of volatile fatty acids, since the pH titration kept the pH inside the reactor around neutral, which was most suitable for methanogenic bacteria. The activity of methanogenic bacteria was inhibited when the environmental conditions were too acid for them. This was indicated by the fatty acids accumulation of fatty acids and a decreasing pH. By adjusting the pH around neutral, the methanogenic bacteria were capable of converting acetate at maximum rates, so they could prevent the very high accumulation of fatty acid that was observed without pH titration. However more acetate was produced at 4 d HRT than could be converted by Methanosarcina and Methanosaeta at optimal conditions.

The pH regulation system is required when treating whey at high OLR. It will maintain the optimum pH condition for methanogenesis. At high OLR, more fatty acids were produced than could be degraded and this determined the environmental conditions inside the reactor. A decrease in pH can retard the conversion process of fatty acids to biogas, which will lead to a anaerobic process failure. In this experiment, the pH regulation system was essential for treating wastewater with OLR higher than 10 g $COD \cdot l^{-1} \cdot d^{-1}$.

3.3 Batch Assay with Suspension Culture from AnFB Reactor for Treating Various Substrates

Batch assay experiments were done in order to find out the capability of the mixed enrichment culture that grew in the reactor for converting precursor substrates in the methanogenesis process, in this case, H_2/CO_2, formate, methanol and acetate. The batch assay experiments were made in a serum flask with a volume of 110 ml. The inoculum for the test was the effluent suspension which was taken from AnFB reactor and the volume of inoculum was 20% of the total assay volume. The anaerobic conditions were obtained by purging the headspace with nitrogen gas (0 bar) for removing the air, except for experiments with H_2/CO_2 gas. During the observation period, the assays were incubated on an orbital shaker at 37°C with 100 rpm.

Each series of experiments was run with three different concentrations and a control, except for H_2/CO_2. The results will be presented after correction for results of control assay.

The results of batch assays were used for calculating the kinetic parameters of methane production from each substrate. The v_{max} and K_m were calculated using Lineweaver-Burk transformation.

3.3.1 Batch Assay with H_2/CO_2 as Precursor Substrate

Hydrogen and carbon dioxide were used as the substrate for methanogenesis in these experiments. The varying concentrations of the gas were achieved by filling different pressure of H_2/CO_2 (80%:20% gas mixture): 1 bar as the control, 1.25 bar, 1.50 bar, 1.75 bar, 2.00 bar, 2.25 bar and 2.50 bar. That are correspond to 4 mmol, 5 mmol, 6 mmol, 7 mmol, 8 mmol, 9 mmol and 10 mmol. The methane production was measured by gas chromatography daily. The results are presented in Figure 26.

The methanogenic bacteria used hydrogen and carbon dioxide to produce methane. The biochemical reaction follows this equation 39:

Equ. 39: $$4\ H_2 + CO_2 \rightarrow CH_4 + 2\ H_2O$$

In this process one mole hydrogen will produce 5.6 litre methane under STP condition.

All inocula showed positive results. Methane production was detected only in a few hour after supplying the substrate. The total methane production was close to the theoretical amount per mole hydrogen consumed. The production rate of methane was decreasing with the increase of the pressure of hydrogen gas in the gas phase or the concentration of hydrogen gas in the liquid medium. Hydrogen with a pressure of less than 2.25 bar was converted faster than at higher pressure. At higher pressure (greater than 2.25 bar) methane production was delayed and occurred at about the same rate as seen for lower hydrogen concentrations.

This showed that the suspension culture in the reactor was capable of converting hydrogen and carbon dioxide to methane. It meant that hydrogen-utilizing methanogenic bacteria were present in the microorganisms population. The presence of this group with high activity in the whey reactor would maintain the hydrogen partial pressure low, which allowed the conversion process of butyrate to acetate and hydrogen by the respective acetogenic population.

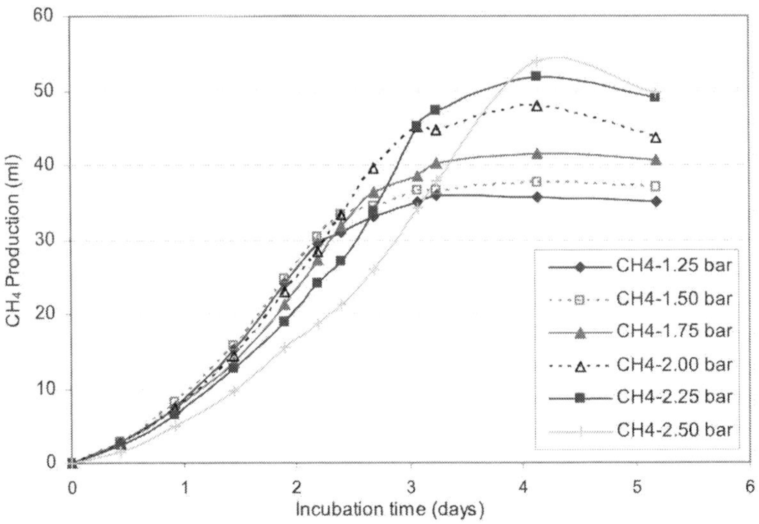

Figure 26 Methane production in batch assays with different amounts of gas mixtures (H_2/CO_2) as substrates.

The kinetics of methane production from H_2/CO_2 gas was calculated from the experimental results. The calculations are listed in Table 11. Higher concentrations of the gas mixture up to 9 mmol (2.25 bar) increased the methane production rate, after that the methane production rate was declining.

Table 11 The kinetic parameters of methane production in batch assays with different amounts of gas mixture (H_2/CO_2) as substrates.

Gas conc. (bar)	Methane production rate ($mmol \cdot l^{-1} \cdot h^{-1}$)
1.25	1.751
1.50	1.860
1.75	1.880
2.00	1.911
2.25	2.126
2.50	1.984
V_{max} ($mmol \cdot l^{-1} \cdot h^{-1}$)	**K_m ($mmol \cdot l^{-1}$)**
2.195	0.26

3.3.2 Batch Assay with Formate as Precursor Substrate for Methanogenesis

Formate is another substrate for methanogenesis in wastewater. It is not always a direct precursor of methane since it is first converted to carbon dioxide and hydrogen. This compound is hydrolyzed by formate dehydrogenase which generates hydrogen and carbon dioxide. Further conversion is similar with the process for hydrogen and carbon dioxide gas. The reactions are as follows (equation 40, 41 and 42):

Equ. 40: $4\ HCOOH \rightarrow 4\ H_2 + 4\ CO_2$

Equ. 41: $4H_2 + CO_2 \rightarrow CH_4 + 2\ H_2O$

Equ. 42: sum: $4\ HCOOH \rightarrow CH_4 + 3\ CO_2 + 2\ H_2O$

When formate is completely converted to methane, one mole of formate theoretically produce 5.6 litre methane and 16.8 litre carbon dioxide. In the batch assay carbon dioxide production is difficult to measure by only examining the gas composition since some of the carbon dioxide will dissolve into the medium under the carbonate equilibrium.

In this experiment, formate was supplied for the culture as sodium formate (CHOONa), with concentrations of 10 mM, 50 mM and 100 mM . These concentrations were equal to a formate amount in the culture of 0.2 mmol, 1 mmol and 2 mmol. The results from this assay are presented in Figure 27.

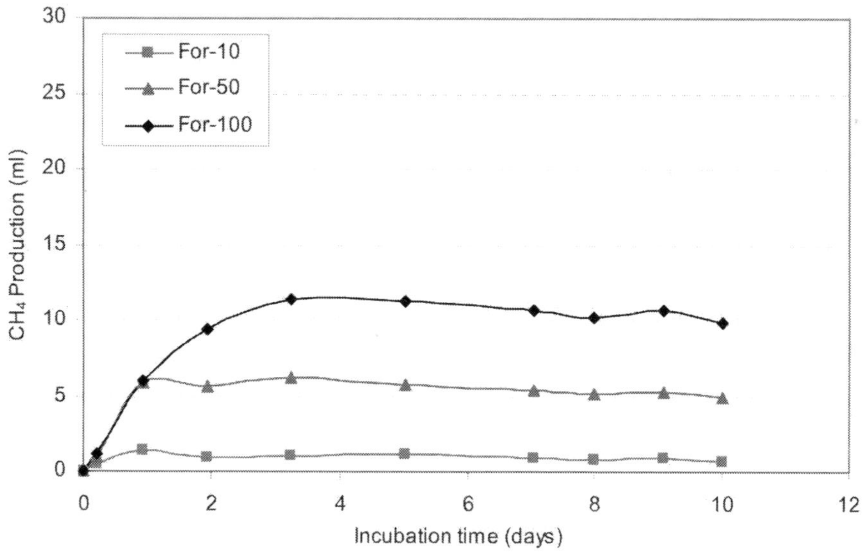

Figure 27 Methane production in batch assay with Na-formate as the carbon source.

All assays gave a positive response of methane production after formate addition. Only in a few hours after the formate addition, methane was produced. It meant that the mixed culture of whey reactor could use formate for methane production. For concentrations of 10 mM and 50 mM the maximum of methane production was reached after the first day, while for 100 mM the maximum of methane production was reached after the third day. The total methane production was increasing with the increase of the formate concentration. The methane production in all assays closed to the theoretical amount. It indicated that the population was capable to use formate for producing methane.

The kinetics of methane production from formate were calculated from the experimental results. Rates and kinetic constants are listed in Table 12. The methane production rates were increasing with increasing formate concentration up to 50 mM. After that a further increase of the substrate did not give any significant effect on methane production rates (Table 12). In wastewater digestion concentration of more than 50-100 mM are not expected, since non-methanogenic bacteria would cleave formate with FDH to CO_2 + H_2O.

Table 12 The kinetics of methane production in batch assay with Na-formate as the carbon source.

Formate conc. (mM)	**Production rate ($mmol \cdot l^{-1} \cdot h^{-1}$)**
10	0.350
50	0.750
100	0.700
V_{max} ($mmol \cdot l^{-1} \cdot h^{-1}$)	**K_m ($mmol \cdot l^{-1}$)**
0.883	15

Utilizing sodium formate did not reduce the production of hydrogen so the amount of methane was the same as for formic acid. It only reduced the amount of carbon dioxide significantly since part of the carbon dioxide reacted with sodium ions to produce sodium carbonate. The pH was slightly more increased after sodium formate was converted to methane and carbon dioxide.

3.3.3 Batch Assay with Methanol as Precursor Substrate

Methanol is a good substrate for some species of methanogenic bacteria including *Methanolobus sp.* and *Methanosarcina sp.* It is a direct precursor substrate for methane production. After induction of methanol dehydrogenase and a special enzyme system, bacteria can convert methanol and hydridocobalamin to methyl-B-12. The formation of methane from CH_3-B_{12} is proceeding via CH_3-CoM and furnished by oxidation of methanol to CO_2. The stochiometries of this fermentation are (equation 43, 44 and 45):

Equ. 43: $CH_3OH + H_2O \rightarrow CO_2 + 6\ H$

Equ. 44: $3\ CH_3OH + 6\ H \rightarrow 3\ CH_4 + 3\ H_2O$

Equ. 45: sum: $4\ CH_3OH + 6\ H \rightarrow 3\ CH_4 + CO_2 + 2\ H_2O$

One mole methanol can be converted to 16.8 litre methane during methanogenesis.

The methanol was added into the culture bottles with concentrations of 10 mM, 50 mM and 100 mM, that corresponded to 0.2 mmol, 1 mmol and 2 mmol of methanol in the medium. The experimental results are presented in Figure 28. In comparison with the first two substrates, the methane fermentation from methanol was relatively slower. It has a lag phase of about two days. After that all assays gave a positive response. The maximum methane production was reached after 4 days for 10 mM methanol, 6 days for 50 mM methanol and 8 days for 100 mM methanol. The methane production was less than the theoretical amount. It was only 90% for 10 mM and 50 mM while for 100 mM it was only 70%. This might be due to more energy that must be dissipated for maintenance of cells the slower cells growth.

The mixed cultures in the assay were able to produce methane from methanol, although it only produced 70-90 % of its theoretical amount. Higher concentration of methanol seemed to have an inhibition effect on the methane fermentation since higher concentrations reduced the maximum methane production and prolonged the lag phase. Since in this experiment there was no addition of minerals and vitamins, the medium might have been deficient of co-factors, which were important for this fermentation.

From the experimental results, the kinetic parameters for methane productions were calculated. They are listed in Table 13. From the results, increasing the concentration of methanol would not be favourable for methanogenesis.

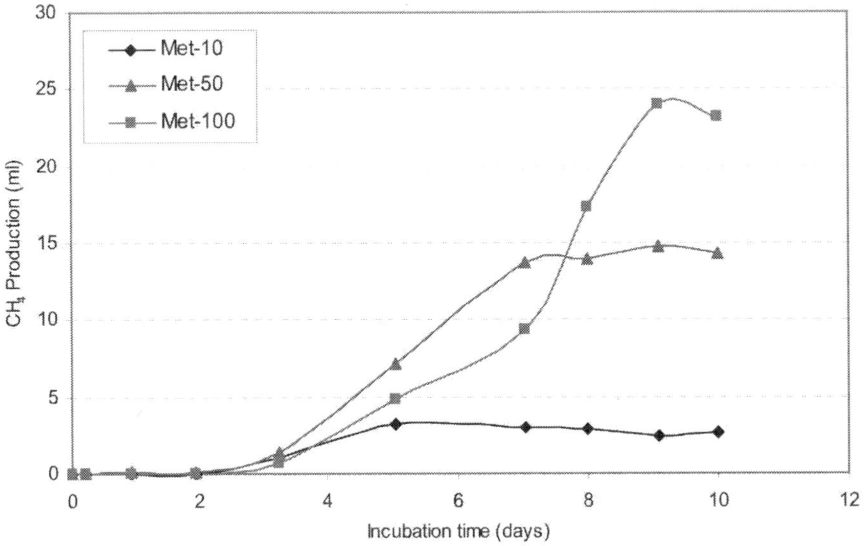

Figure 28 Methane production in batch assay with methanol as the carbon source.

Table 13 The kinetics of methane production in batch assay with methanol as the carbon source.

Methanol conc. (mM)	**Production rate (mmol·l^{-1}·h^{-1})**
10	0.150
50	0.350
100	0.500
V_{max} (mmol·l^{-1}·h^{-1})	**K_m (mmol·l^{-1})**
0.600	30

3.3.4 Batch Assay with Acetate as Precursor Substrate

Acetate is the most important methanogenic substrate in natural environment. It is fermented by methanogenic bacteria to methane and carbon dioxide (equation 46).

Equ. 46: $CH_3COOH \rightarrow CH_4 + CO_2$

One mole of acetate can be converted to 22.4 litre methane and 22.4 litre carbon dioxide.

The acetate was added as sodium acetate into the assays with concentrations of 10 mM, 50 mM and 100 mM, that corresponded to 0.2 mmol, 1 mmol and 2 mmol of acetate in the medium. The experimental results are presented in Figure 29. This assay gave an unexpected result since the methane production was very little. The result was far from the theoretical amount of methane production that would have been 4.48 ml, 22.4 ml and 44.8 ml.

During the observation period, in this case 12 days, the methane production was only less then 10% of its theoretical amount. It was very surprising since the culture in the reactor, which was the source of the inoculums, had the capability for degrading the accumulated acetate in overloaded reactor immediately, when overloading was stopped. The difference must be due to the source of inoculums. For batch assays, the microorganisms came from the effluent of the whey reactor. Apparently most of the acetate converting methanogens grew attached to the supporting material in the biofilm.

Acetate-utilizing methanogenic bacteria seemed to be unable to grow as a single-cell system. This was in agreement with the statement of Vogels et al. (1988) that acetate conversion has been hampered by the lack of suitable cell systems. Most of the successful studies about acetate conversion used a particulate system.

Growth systems with aggregated cells or cells that were attached to substrate provide a suitable environment for growth of acetate-utilizing methanogens such as low redox potential and anaerobic conditions. This could explain the failure of acetate conversion in the batch assay experiments.

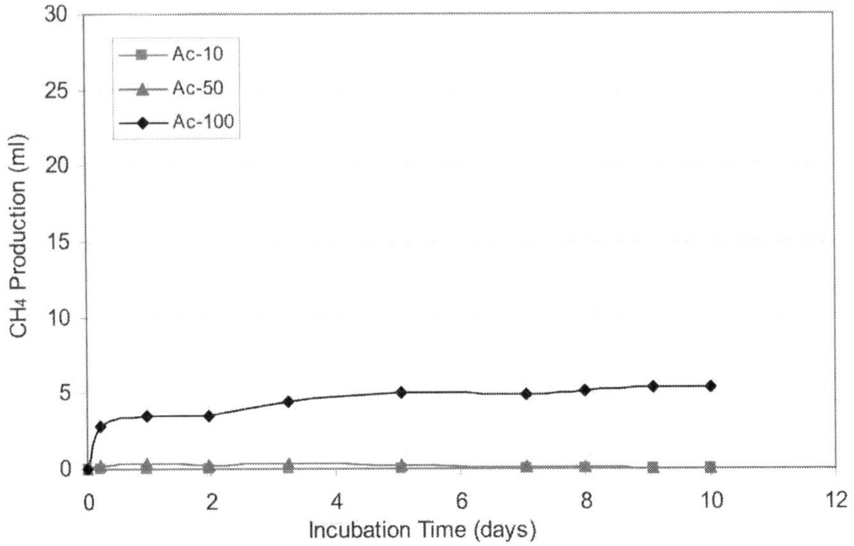

Figure 29 Methane production in batch assay with Na-acetate as the carbon source.

3.3.4.1 Acetate as Precursor Substrate and Addition of a Reducing Agent

In the previous experiment, the population in the reactor effluent was unable to convert acetate to methane. It seemed that a low enough redox potential, which was required by these strict anaerobes, could not be maintained in the biofilm. Methanogenic bacteria grow only in media with a redox potential lower than E'_0 = -0.3 V (Gottschalk, 1979). A low redox potential can be maintained by supplementing the medium with SH-compounds. SH-compounds, such as thioglycolate, cystein, sodium sulfide or resazurin, can lower the redox potential in the whey medium if added as reducing agents.

A SH-compound, in this case sodium sulfide at a concentration of 0.05 $g \cdot l^{-1}$ and resazurin as a redox indicator at a concentration of 0.001 $g \cdot l^{-1}$ were supplemented to the whey inoculum to provide a low redox potential in the whey effluent medium, used for the following experiment. The substrate in this experiment was sodium acetate at concentrations of 0 (= control), 10, 50 and 100 mM. The methane production in this assay is presented in Figure 30.

This assay exhibited a better result than the previous experiment. The population produced methane at a concentration that was higher than the theoretical amount. However, the lag phase in this assay was relatively long, in the range of 30-80 days. A higher concentration of acetate required a longer lag phase.

From the experimental results, the kinetic parameters for acetate conversion were calculated using the Michaelis-Menten model. The kinetic parameters for methane production from acetate are listed in Table 14.

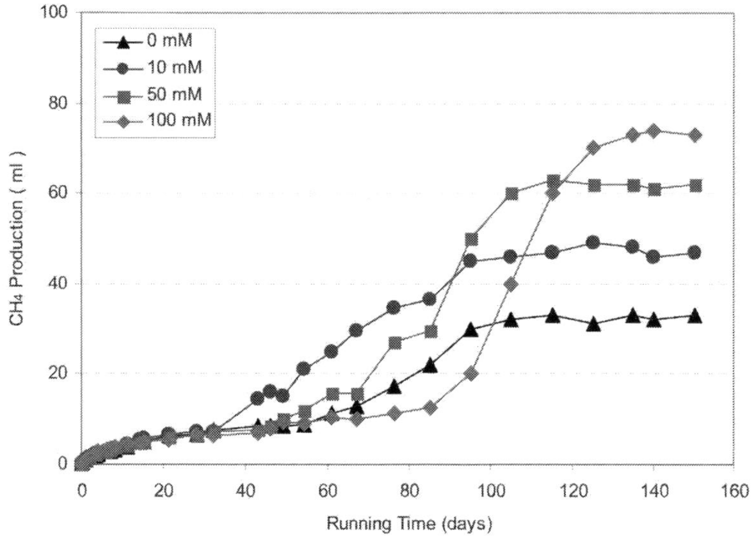

Figure 30 Methane production in batch assays with acetate as the carbon source in the presence of a reducing agent (Na_2S and Resazurin).

Table 14 The kinetic parameters of methane production in batch assay with acetate as the carbon source in the presence of a reducing agent (Na_2S and Resazurin).

Acetate conc. (mM)	**Production rate ($mmol \cdot l^{-1} \cdot h^{-1}$)**
10	0.026
50	0.042
100	0.056
V_{max} ($mmol \cdot l^{-1} \cdot h^{-1}$)	**K_m ($mmol \cdot l^{-1}$)**
0.057	13

The concentrations of volatile fatty acids in the assay were plotted in the graph in Figure 31. As the medium was the effluent of the whey reactor, the volatile fatty acids were not only acetate, which was added as sodium acetate but also some other residual volatile fatty acids that were present at much lower concentrations as intermediate metabolites in the effluent. Propionate, butyrate and other intermediate metabolites were degraded in parallel with the acetate during incubation. Consequently, the methane production was higher than the theoretical amount, which should be produced from the added sodium acetate.

During the lag phase, mainly acetogenesis and some methanogenesis took place. The methanogenesis was carried by methanogenic bacteria which could use H_2/CO_2 and other methanogenic substrates, such as formate, methanol. Acetate was, if converted to methane at all, only slowly utilized as indicated by the increase of the concentration of acetate. After acetogenesis was completed at the day 15 to 30, the acetate was degraded by methanogenic bacteria. During this process, the butyrate and propionate were also degraded together with acetate. Since the population in the effluent of the whey reactor contained hydrogen-utilizing methanogenic, the hydrogen partial pressure was kept low. This was a favourable condition for propionate and butyrate degradation.

The amounts of acetate and methane in the assay are presented in Figure 32. This picture showed clearly that from 1 mole of acetate more than 1 mole of methane was produced, including some other sources for methane production. The exponential phase in methane production coupled with the exponential phase in acetate degradation. In general, the exponential phase in methane production started when the acetate had been degraded around 25% of its maximal concentration.

This experiment demonstrated that a low redox potential was a perquisite for the growth of acetate-utilizing methanogenic in a cell-free system. Acetate-utilizing methanogenic bacteria required a longer lag phase in methane production.

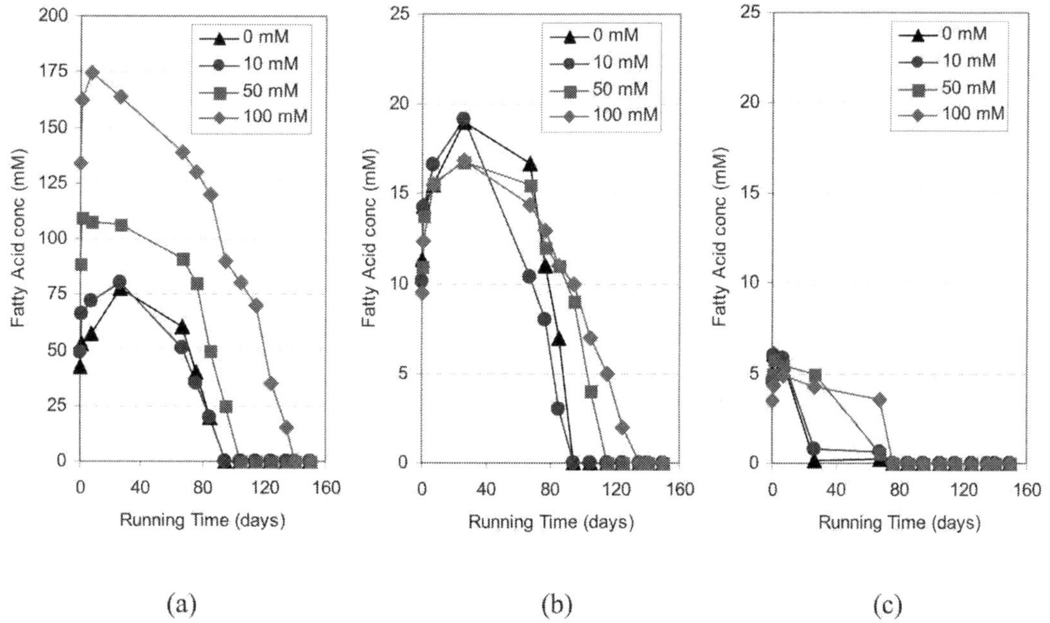

Figure 31 Concentration of VFAs in batch assays with acetate as the carbon source in the presence of a reducing agent (Na_2S and Resazurin); (a) Acetate, (b) Propionate and (c) n-Butyrate.

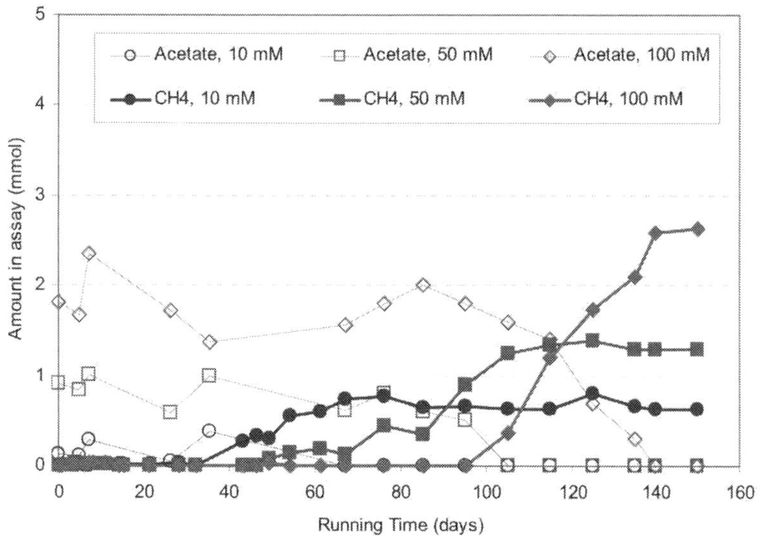

Figure 32 Amount of acetate and methane in each assay (volume of assay = 20 ml per bottle) with acetate as the carbon source in the presence of a reducing agent (Na_2S and Resazurin).

3.3.4.2 *Acetate as Precursor Substrate and Addition of a Reducing Agent together with Trace minerals and vitamins*

Except for requiring a very low redox potential, growth of methanogenic bacteria was stimulated by trace minerals and vitamins solution. In this experiment, the effect of mineral and vitamins addition on methane production from acetate was investigated. Trace mineral and vitamin solutions as described in Medium 141 (DSMZ) were used. The results of methane production are presented in Figure 33.

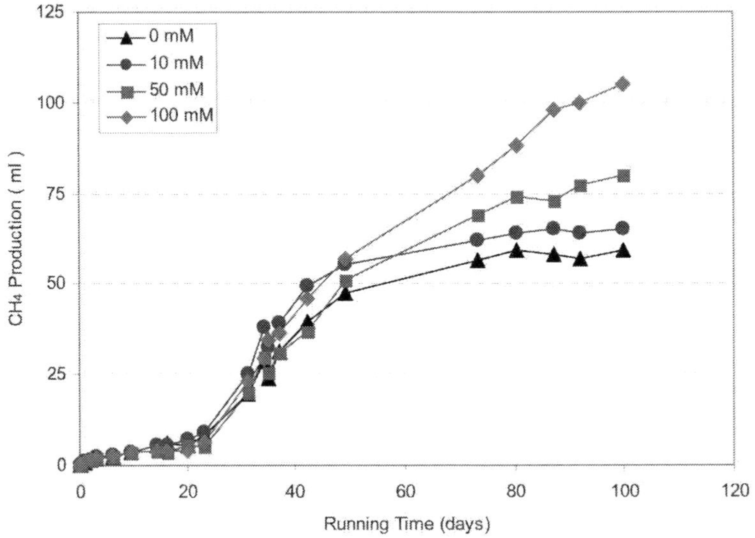

Figure 33 Methane production in batch assays with acetate as the carbon source in the presence of a reducing agent (Na_2S and Resazurin) and with addition of trace minerals and vitamins.

This experiment showed better results than the previous experiment with the addition of only the reducing agent. The trace minerals and vitamins improved the methanogenesis by shortening the lag phase and increased the total methane production. In this assay, the initial concentration of acetate had no significant effect on the length of the lag phase. Different amounts of acetate in the assay required a similar lag phase period, of only 20 days for methanogenesis to start. This indicated that the length of lag phase was not determined by the concentration of the substrate but by the supply of other elements which improved the metabolism process. The lag phase in this experiment was 10 to 60 days shorter than in the assay without trace minerals and vitamins.

The kinetic parameters of methane production from acetate were calculated and tabulated in Table 15. The V_{max} of methane production with trace mineral and vitamin supplementation was higher than in assays without minerals and vitamins. A higher maximum methane production rate indicated that the population in this assay was more active or that the population size had increased.

Table 15 Kinetic parameters of methane production in batch assays with acetate as the carbon source in presence of a reducing agent (Na_2S and Resazurin) and with addition of trace minerals and vitamins.

Acetate conc. (mM)	Production rate ($mmol \cdot l^{-1} \cdot h^{-1}$)
10	0.018
50	0.039
100	0.056
V_{max} ($mmol \cdot l^{-1} \cdot h^{-1}$)	**K_m ($mmol \cdot l^{-1}$)**
0.065	27

The concentrations of volatile fatty acids in the assay were presented in Figure 34. During the lag phase (the first 20 days), a small amount of acetate was produced. This could be from acetogenesis of intermediate metabolites. Propionate only contributed of acetate concentration after the lag phase, since the propionate degradation started after day 60. During the first 40 days, there was no significant change in propionate because the degradation of butyrate was proceeding during this period. The butyrate was converted to acetate and hydrogen and prevented propionate degradation. The degradation of volatile fatty acids was influenced by the presence of trace minerals and vitamins. These shortened the degradation period for each fatty acid.

The total amount of acetate and methane in the assay is presented in Figure 35. Methane was stoichiometrically produced from acetate. This experiment showed that trace minerals and vitamins reduced the time of the lag phase in methane production and reduced the period of volatile fatty acid degradation.

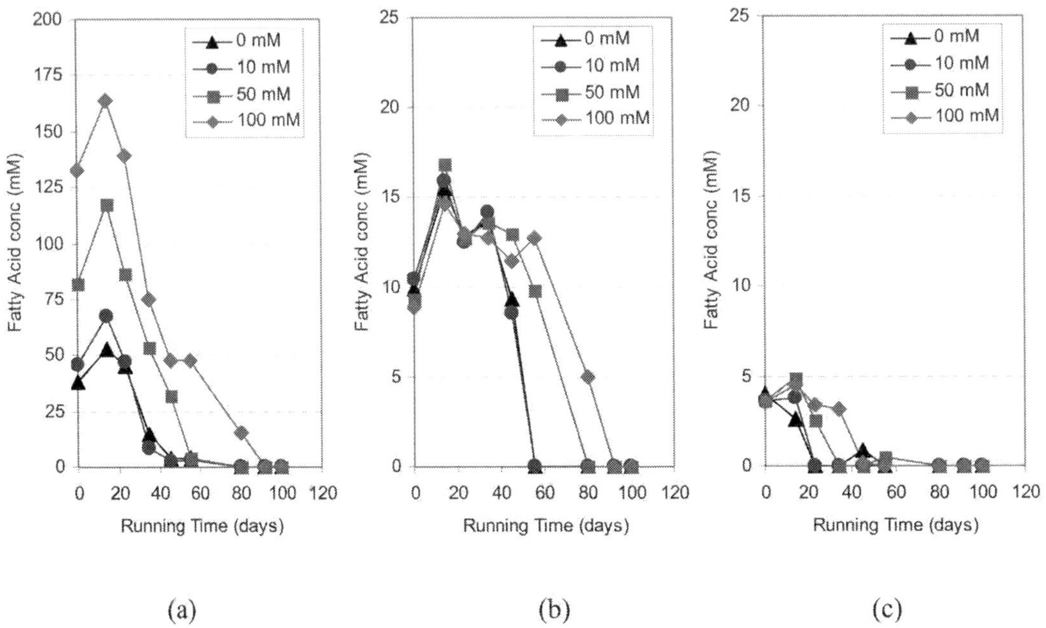

Figure 34 Concentration of VFAs in batch assay with acetate as the carbon source in the presence of a reducing agent (Na_2S and Resazurin) and with the addition of trace minerals and vitamins; (a) Acetate, (b) Propionate and (c) n-Butyrate.

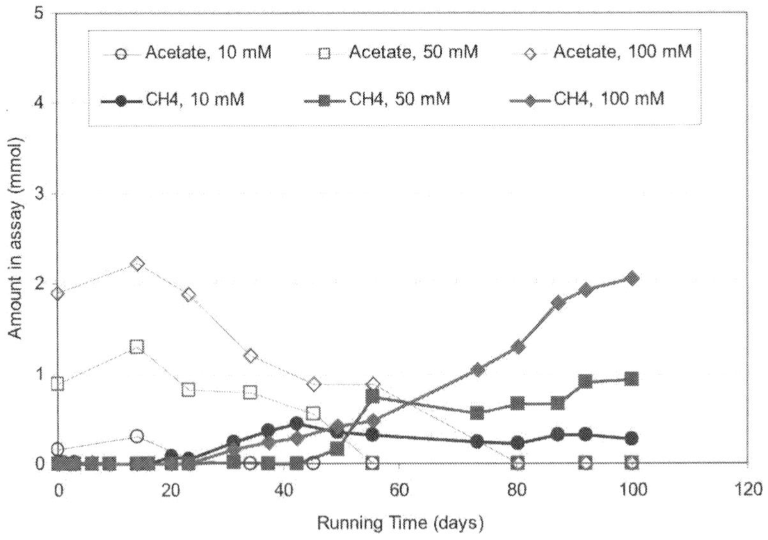

Figure 35 Amount of total acetate and methane in each assay (volume of assay = 20 ml per bottle) with acetate as the carbon source in the presence of a reducing agent (Na_2S and Resazurin) and in the presence of trace minerals and vitamins.

3.3.5 Batch Assay with Residual VFAs in the Effluent of the Whey Reactor. Acetate as Precursor Substrate for Methanogenesis and the Effect of Minerals and Sulfate Addition

3.3.5.1 Effect of Minerals

In the previous experiment, the suspended population in the culture was capable to use acetate as a precursor substrate for methanogenesis after reducing agents, trace minerals and vitamins were added into the assays. This showed that trace minerals had a positive effect on improving methanogenesis. In order to determine the essential mineral for methanogenesis, some minerals, in this case iron (Fe^{3+}), nickel (Ni^{2+}), cobalt (Co^{2+}), manganese (Mn^{2+}) and magnesium (Mg^{2+}) were added into the culture either as single ions or by combination of several minerals. The final concentration was 50 $mg \cdot l^{-1}$ in the assay for each mineral. The substrate in the assay was the remaining acetate in the effluent of AnFB reactor. The initial concentration of acetate in the assay was 40 $mmol \cdot l^{-1}$. The assays were incubated on an orbital shaker at 37°C under N_2 gas on the gas space. The results are presented in Figure 30.

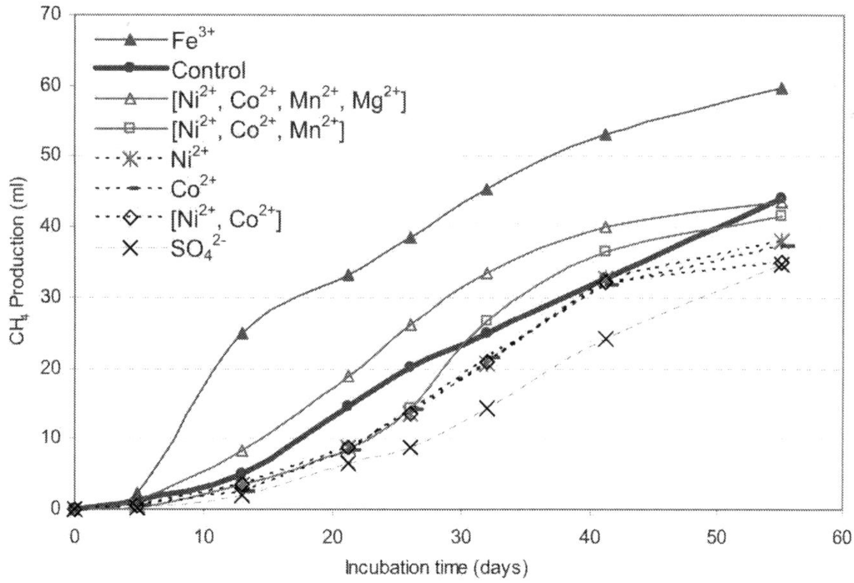

Figure 36 Dependence of methane production in batch assays with residual VFAs in the effluent of whey reactor on minerals addition.

In this experiment there were seven variations of mineral addition and one control without mineral addition. When methane production was used as a parameter of methanogenesis, there were only three of seven variations with a positive effect. This meant that in these assays methane production was higher or close to equal with the result of the non-minerals supplemented control. The exponential phase of methane production in these assays started after 5-15 days for all variations even though the methane production rates were not similar. The methane production reached its highest value after 55 days.

The mineral additions which had a positive effect were Fe^{3+}, a combination of Ni^{2+}, Co^{2+}, Mn^{2+} and Mg^{2+} and a combination of Ni^{2+}, Co^{2+} and Mn^{2+}. Only in the assay with Fe^{3+} methane production was significantly higher than the amount that was produced by the control. The other assays with mineral additions produced little less a similar amount of methane as the control. Ferric ions improved methanogenesis more than the others minerals. Additional of ferric ions into the culture increased the total methane production by 30%, compared to the control. Ferric ions also shortened the lag phase of this culture. While comparing the two assays with a combination of minerals, the addition of Mg^{2+} to a mineral combination of Ni^{2+}, Co^{2+} and Mn^{2+} enhanced the methanogenesis which was indicated by a faster and better total methane production.

Trace amounts of nickel and cobalt were known as co-factors for methanogenesis. However, their addition in this experiment as a single mineral solution gave no positive result. In contrary, the methane production was less than that of the control. The total methane production with addition of Ni^{2+} or Co^{2+} was only 80% of the control. It might be that the concentration of these elements in the whey was already high and further addition increase the concentration close to the concentration that might inhibit methanogenesis. $NiCl_2$ at 60 $mg \cdot l^{-1}$ revealed 50% inhibition of methanogenesis in a whey-containing assay (Zayed and Winter, 1998).

Combining Ni^{2+} and Co^{2+} was not improving methanogenesis in the assay. It even inhibited methane production to 77% of the control. The effect of the presence of these ions in combination increased the inhibitory effect over that was observed with either single ion.

Addition of manganese ions to the combination of nickel and cobalt had a positive result on methane production. The methane production reached approximately that of the control. Although the total methane production was in the order of the volume of methane produced by the control, it had a higher production rate. Manganese ions apparently improved methanogenesis. Manganese ions seemed to reduce the inhibitory effect of nickel and cobalt in the assay.

Magnesium ions were added to the combination of nickel, cobalt and manganese ions. This ion combination increased the production of methane. The total methane production was almost equal to the production of the control. The presence of magnesium ions at a concentration near the stimulatory range, of 75-150 $mg \cdot l^{-1}$ (Grady and Lim, 1980), would ensure optimum metabolic activity of bacteria under normal condition and also would help reduce toxic effects.

Figure 37 presents the amount of acetate, methane and carbon dioxide gas in each assay with additional ferric ions, combinations of nickel, cobalt, manganese and magnesium and the control. In all assays, the increasing methane production was coupled to decreasing acetate concentrations.

In every assay, acetogenesis was detected in the first 5-12 days as the concentration of acetate increased 20-50% compared to the initial concentration. The effect of minerals on improving acetic acid cleavage to methane and CO_2 was uneasy to determine since the maximum concentrations of acetate in this assays were less than those of the control. This might be due to a more rapid conversion of acetate to methane in the presence of minerals. The increased methane production in assays with minerals indicated indirectly that acetogenesis must have been improved by mineral addition since acetate was the precursor substrate for methanogenesis.

The rates of acetate conversion and methane production are listed in Table 16. Ferric ion gave the best results for acetate conversion and methane production. The rates were about 2.5 times higher than those of the control. The combination of minerals (Ni, Co, Mn and Mg) did not favour acetate conversion and methane production.

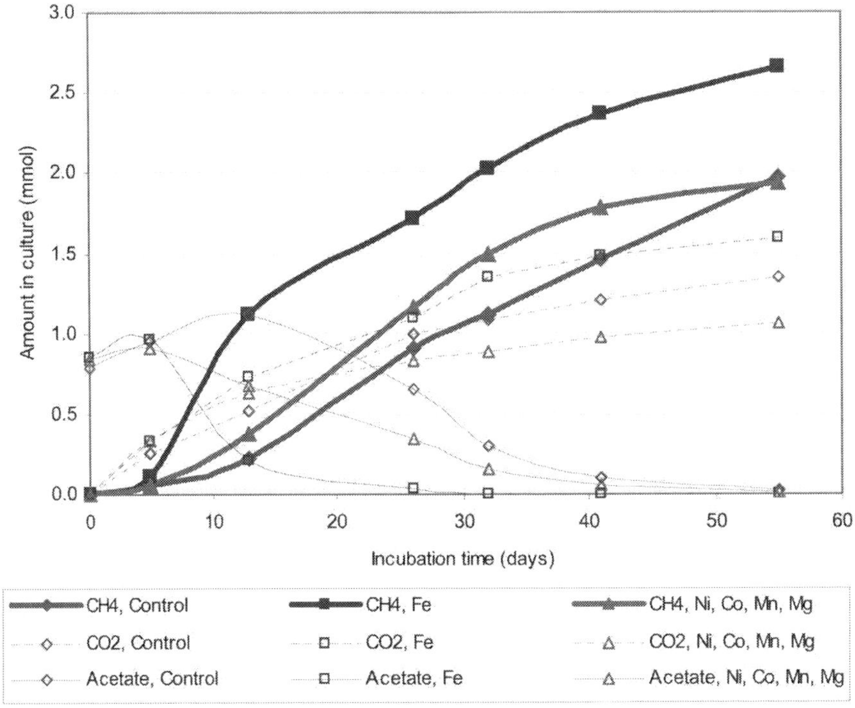

Figure 37 Amount of acetate, methane and carbon dioxide in each assay (volume of assay = 20 ml per bottle) that contained a combination of cations.

Methane production rates were apparently higher than acetate conversion rates. This was due to the fact that acetate was not the only substrate in the assay. Since the effluent of the AnFB reactor was used as an inoculums, other intermediate metabolites, volatile fatty acids, formate, and hydrogen and carbon dioxide might have been additionally available substrates for methanogenesis. This can be seen from Figure 37. After the acetate had been used up, some methane was still produced. During this phase the bacteria used other substrates than acetate. As a result, methane production rates were around 40-70% higher than acetate conversion rates. This experiment proved that ferric ions were suitable minerals for improving methanogenesis from a whey-based wastewater.

Table 16 Rates of acetate conversion and methane production with mineral addition (concentration of each mineral = 50 $mg \cdot l^{-1}$).

Variation	Rate ($mmol \cdot l^{-1} \cdot h^{-1}$)	
	Acetate conversion	**Methane production**
Control	- 0.079	0.100
Fe^{3+}	- 0.196	0.280
Ni, Co, Mn, Mg	- 0.063	0.110

3.3.5.2 Effect of Sulfate Anions

Sulfate is the major anion which often present in wastewater. Sulfate was added to the whey medium of the assay (Figure 36). Methane production was compared to that of the control. This experiment showed that the methane production in the assay with sulfate addition was less than that without sulfate addition (=control). It was only 75% of the control. Methanogenesis seemed to be retarded in the presence of sulfate. Since the anaerobic mixed culture might contain some sulfate reducing bacteria, there might be a competition between methane fermentation and the sulfate reducing process when sulfate is available. Without sulfate, the reducing equivalents from the organic substrate appeared as methane, whereas in the presence of sulfate they were required for hydrogen sulfide production. The inhibition of methanogenesis could be a result of competition for the common hydrogen substrate if methanogenesis and sulfate reduction were alternative electron acceptor . As long as sulfate was present, electron flow into sulfate reduction would be the favoured reaction (Widdel, 1998). This result proved that the whey inocula contained sulfate reducing bacteria that competed with the methanogenic bacteria in using hydrogen or reduced organic substances. As a result, the methane production was less than that of the control.

With an initial concentration of acetate of 40 $mmol \cdot l^{-1}$ in the assay, methane production was expected around 17.9 ml. The results showed that the volume of methane was at least twice of the theoretical value. Since in the culture, acetate was not the only available substrate for methanogenesis, the volume of methane production was higher that the theoretical amount which might be produced from acetate alone. A part of methane was produced from acetate while the rest could have been from other intermediate products or hydrogen and carbon dioxide.

3.4 Implementation of ferric ion addition

3.4.1 Implementing the Mineral Addition in a Batch System for Degrading Residual VFAs in the Effluent of Whey Reactor

In section 3.3.5, ferric ions were shown as essential minerals in degrading residual fatty acids in whey effluent. This experiment tried to reproduce the previous results coupled with supplementation of a reducing agent, in this case sodium sulfide and resazurin. The concentration of ferric ions was 50 $mg \cdot l^{-1}$. The result of methane production is shown in Figure 38. The methane production in this experiment was four times higher than theoretically expected. The additional of ferric ions improved methanogenesis. It increased the methane production during the exponential phase, although the final volume of methane was similar in these two variations.

Since the methane production was much higher than the concentration of acetate which was available in the assay, the methane was not only produced from acetate. Other methanogenic substrate must have contributed to the final methane production.

The concentration of volatile fatty acids in the assay can be seen in Figure 39. The degradation of acetate was coupled with propionate and butyrate production and subsequent degradation. During the first five days, the concentration of acetate increased in the two assays without and with ferric ions. At the same time, propionate and butyrate were formed and there was no difference between the two assays. Butyrate degradation was observed together with acetate degradation, whereas propionate degradation started only later, but then at a high rate. It seemed that ferric ions had stimulatory effect on the acetate degrading methanogenic bacteria.

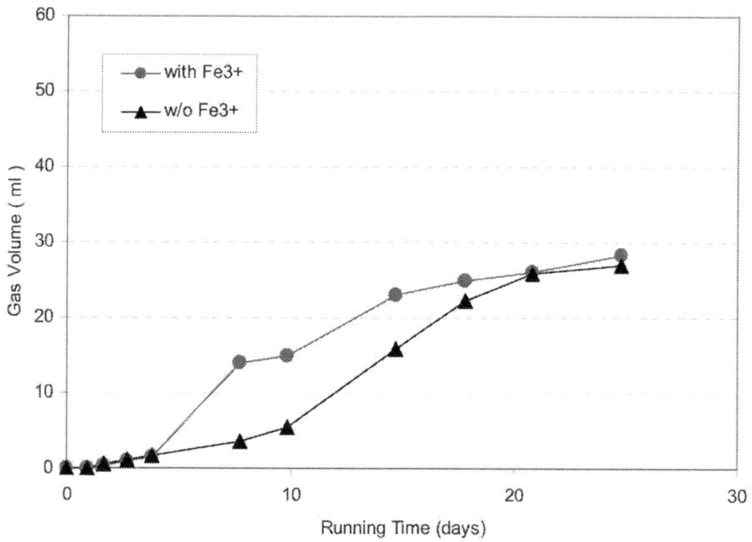

Figure 38 Methane production in batch assay with residual VFAs in the effluent of whey reactor and Fe^{3+} (50 $mg \cdot l^{-1}$) addition.

Figure 40 shows the amount of acetate and methane in both assays. It can be seen clearly that ferric ions had a stimulatory effect on acetate degradation. With the same initial amount of acetate, the assay with ferric ions was always contained less acetate than the assay without ferric ion addition. The acetate conversion rate and the methane production rate of the assay with ferric ions were higher than in the assay without ferric ions. These results were concurrent with the results in section 3.3.5.1.

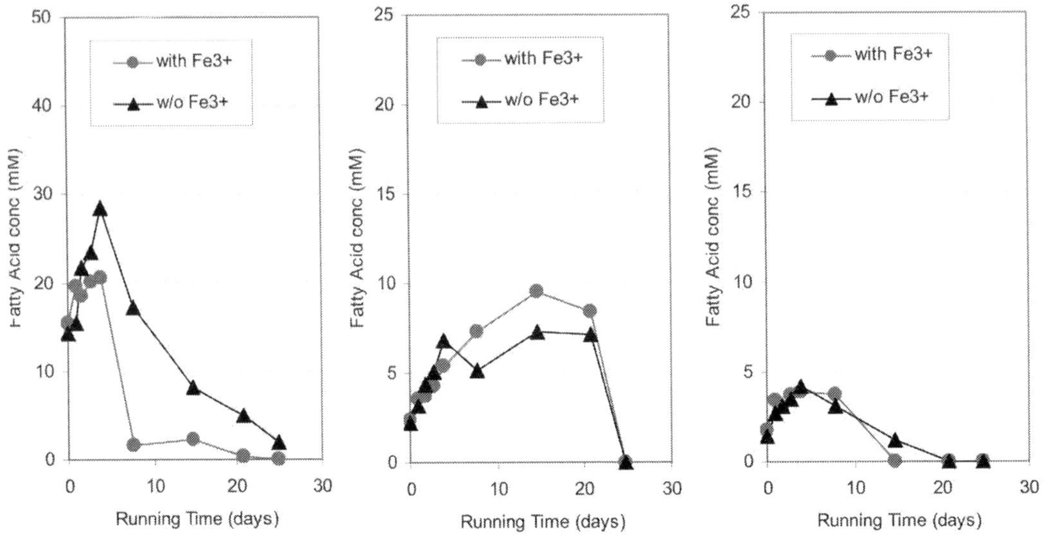

(a) (b) (c)

Figure 39 Concentration of VFAs in batch assay with residual VFAs in the effluent of the whey reactor in the presence of a reducing agent (Na_2S and Resazurin) and in the presence of Fe^{3+} (50 $mg \cdot l^{-1}$); (a) Acetate, (b) Propionate and (c) n-Butyrate.

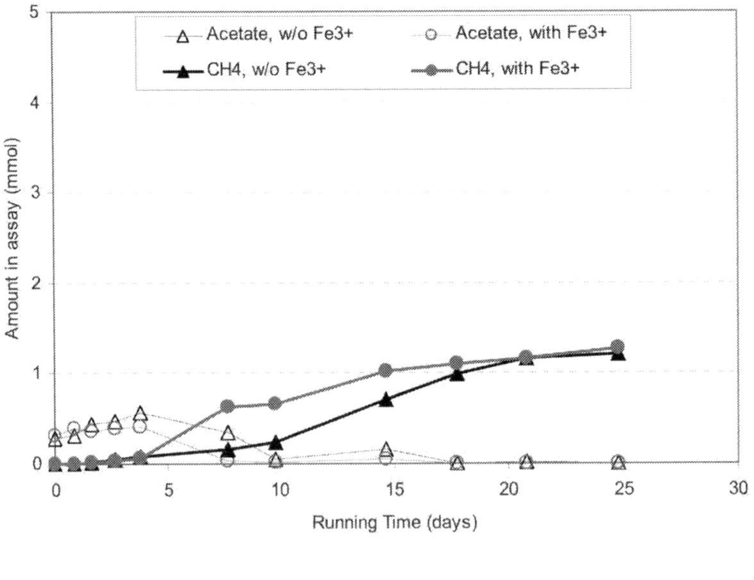

Figure 40 Amount of acetate and methane in each assay (volume of assay = 20 ml per bottle) with Fe^{3+} (50 $mg \cdot l^{-1}$) addition.

Table 17 The rates of acetate conversion and methane production with Fe^{3+} (50 mg·l^{-1}) addition.

Variation	**Rate (mmol·l^{-1}·h^{-1})**	
	Acetate conversion	**Methane production**
Control	- 0.078	0.174
Fe^{3+}	- 0.203	0.278

Ferric ions addition seemed to have an improving effect on methanogenesis by increasing the methane production and acetate conversion rate.

3.4.2 Implementing the Mineral Addition in a Continuous Whey Reactor System

Since ferric ions could improve methanogenesis in a batch system, the addition of ferric ions was implemented in a continuous system in an AnFB reactor at a HRT of 4 days with a whey feeding of 50 g whey·l^{-1}. The mineral that was added to the reactor was ferric trichloride ($FeCl_3$), technical grade. There were four variations in the concentrations of $FeCl_3$, 0 mg·l^{-1} (as control), 10 mg·l^{-1}, 20 mg·l^{-1} and 50 mg·l^{-1}.The $FeCl_3$ was added for the first time when there was no accumulation of volatile fatty acids in the reactor and the pH inside the reactor was around 6.8 (steady state conditions). These conditions were reached after the reactor was switched into batch mode. Once the mineral solution had been added, the reactor was operated in a continuous mode and fed with whey solution. The methane production and pH inside the reactor were measured for evaluating the performance of methanogenesis in the AnFB reactor. The results can be seen in Figure 41.

The experiment with no $FeCl_3$ added, reached 27.5 l·d^{-1} biogas production within one week and the pH was 5.4. The concentration of acetate finally was 40 mM. The fastest decrease of the pH happened on the first day, where the pH dropped 0.8 units.

In the second experiment, the reactor was supplemented with 10 mg·l^{-1} of $FeCl_3$. Within one week, the biogas production was 26 l·d^{-1} and the pH were 5.6. The final concentration of acetate was 35 mM. A fast decrease of the pH occurred in the first two days. The pH dropped 0.7 unit in the first day.

In the third experiment the reactor was supplemented with 20 $mg \cdot l^{-1}$ of $FeCl_3$. It reached 25 l biogas production per day and the pH dropped to 5.5 within one week. The final concentrations of acetate were 40 mM. The pH in the first day had dropped by 1 unit.

In the last experiment, the reactor was supplemented with 50 $mg \cdot l^{-1}$ of $FeCl_3$. The biogas production was only 22 $l \cdot d^{-1}$, but the pH dropped only to 6.0. The final concentration of acetate was 10 mM. Within the first day the pH dropped by 0.3 units.

Increasing the concentrations of $FeCl_3$ in the continuous whey reactor reduced the biogas production. Higher concentrations of $FeCl_3$ led to a lower acetate concentration and as a result the pH inside the reactor decreased only slowly. In general higher concentrations of acetate would give a lower pH. Unexpectedly, in the third experiment, the same concentration of acetate as in the second experiment gave a different pH value. This was due to the fact that in the third experiment the concentrations of propionate and butyrate were high, 12 mM and 5 mM, respectively. It seemed that higher concentrations of $FeCl_3$ retarded the acetate production but favoured the accumulation of the other volatile fatty acids. It was shown clearly that at a concentration of 50 $mg \cdot l^{-1}$ Fe^{3+}, at which the concentrations of acetate was only 10 mM. Other volatile fatty acids were not found within the first week. Thus, the $FeCl_3$ had some inhibition effects on acidogenesis and acetogenesis.

As the acetogenesis was inhibited, the acetate as one precursor substrate for methanogenesis became limiting. At the end this situation resulted in less methane production, as shown for Fe^{3+} concentrations of 50 $mg \cdot l^{-1}$. The methane production ceased when the concentrations of acetate decreased. The concentrations of volatile fatty acids in the assay during these experiments are presented in Figure 42.

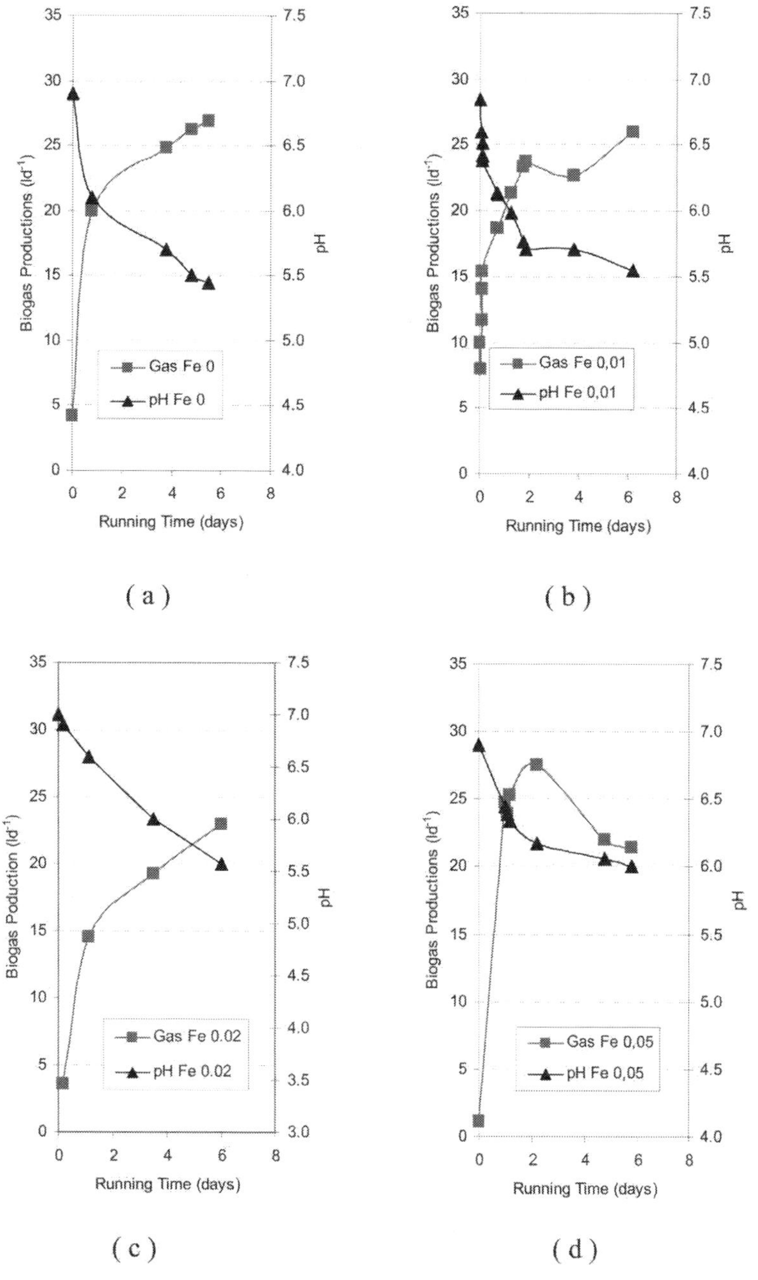

Figure 41 Changes of pH and methane production in the AnFB reactor at a HRT of 4 days. The concentration of the feeding was 50 g whey·l^{-1} with addition of $FeCl_3$; (a) 0 mg·l^{-1} – control, (b) 10 mg·l^{-1}, (c) 20 mg·l^{-1} and (d) 50 mg·l^{-1}.

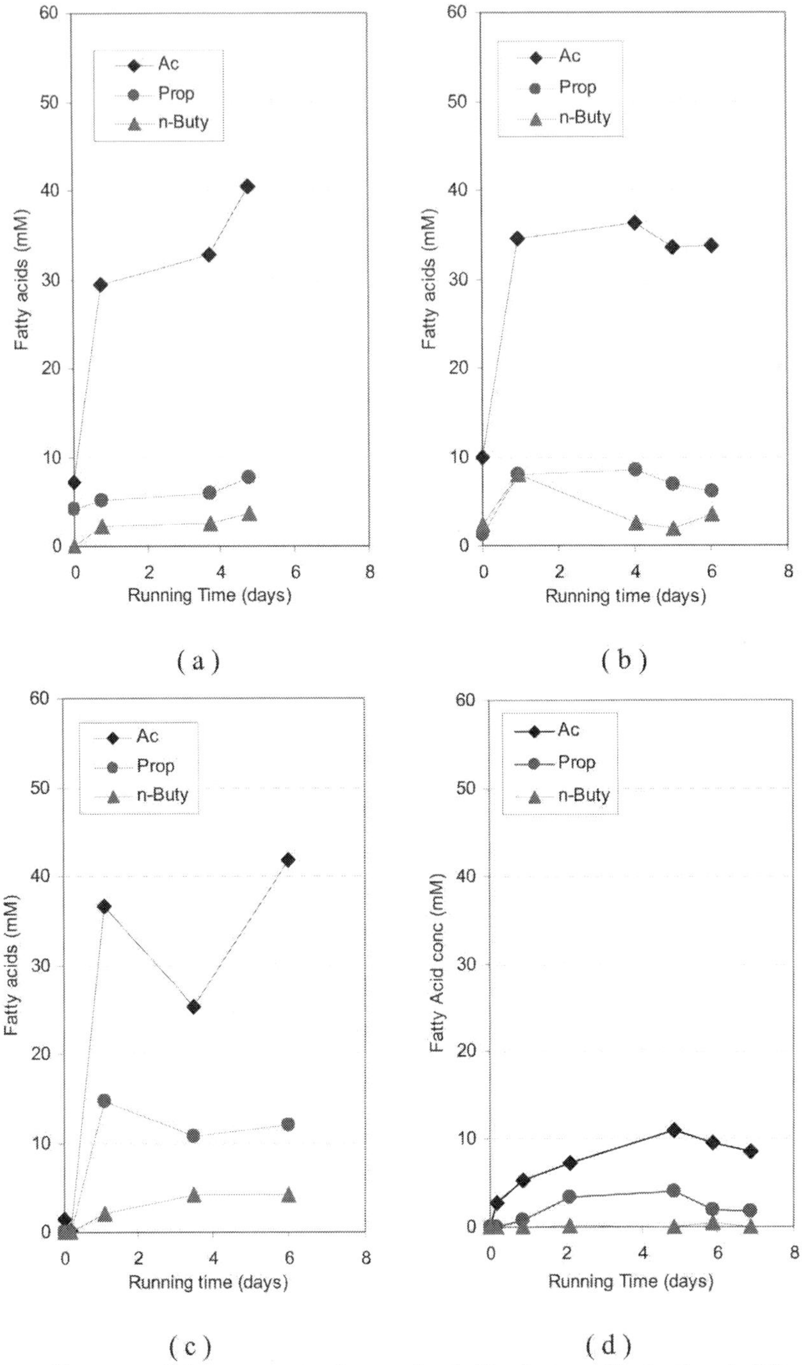

Figure 42 Changes of the concentrations of volatile fatty acids in the AnFB reactor at a HRT of 4 days. The concentration of the feeding was 50 g whey·l^{-1} with addition of $FeCl_3$; (a) 0 mg·l^{-1} – control, (b) 10 mg·l^{-1}, (c) 20 mg·l^{-1} and (d) 50 mg·l^{-1}.

Chapter 4

DISCUSSION

4.1 Conversion of Volatile Fatty Acids (VFAs)

The VFAs conversion process determined the degradation or accumulation of VFAs that were produced in the fermentative phase during anaerobic whey treatment. When there was no new substrate supply for the reactor, the accumulated VFAs were degraded. This operating mode can be used as an experimental option to analyze the prevailing degradation rates for fatty acid intermediates during high-rate whey methanation.

The degradation of a mixture of VFAs in batch or continuous whey fermentation systems had a similar trend. In general, at the beginning of the incubation time for VFAs degradation, the concentration of acetate increased when at the same time the concentration of butyrate decreased. After butyrate was degraded, the concentration of propionate had increased slightly and the concentration of acetate started to decrease. At a low acetate concentration, propionate was also degraded.

The degradation of VFAs was only possible by syntrophic association of the population in the reactor. A syntrophic association is an obligatory cooperation of at least two bacterial species in which both partners depend on each other to perform the metabolic activity which was observed. The mutual dependence of the two species on each other cannot be overcome by simply adding a co-substrate or any other type of nutrient (Schink, 1991) During methanogenic degradation of complex organic matter, different anaerobic bacteria must cooperate. For fatty acid degradation of acetogenic and methanogenic bacteria must be allowed based on interspecies hydrogen transfer.

During degradation of whey, the spectrum of VFAs consisted mainly of acetate, propionate and n-butyrate. Sometimes n-valerate and branched-chain fatty acids, such isobutyrate and iso-valerate were also found in the effluent, although their concentrations were insignificant compared to the three main fatty acids. The branched-chain fatty acids are formed during fermentative degradation of the corresponding amino acids by oxidative deamination and decarboxylation (McInerney, 1988). Branched-chain fatty acid formation might proceed during anaerobic degradation of whey, which contained around 12% protein. All fatty acid-oxidizing syntrophs known so far are even more specialized and can use fatty acids only (Schink, 1991).

In methanogenic bacteria e.g. *Methnosarcina barkeri* or *Methanothrix soehngenii*, acetate is usually cleaved into methane and carbon dioxide by decarboxylation. (equation 47) (Schink, 1990). The reaction of acetate oxidation that was stated by Zeikus (1977) could supply energy to a bacterium's energy metabolism.

Equ. 47: $CH_3COOH \rightarrow CH_4 + CO_2$ ($\Delta G_0' = -36$ kJ/mol acetate)

There is only one defined culture, *Syntrophobacter wolinii*, that synthropically oxidizes propionate to acetate and CO_2 (Schink 1991). The reaction of propionate oxidation, both in a ternary and a binary mixed culture are (equation 48, 49):

Equ. 48: $4CH_3CH_2COO^- + 4H^+ + 2H_2O \rightarrow 7CH_4 + 5CO_2$

($\Delta G_0' = -62.25$ kJ/mol propionate)

Equ. 49: $4CH_3CH_2COOH + 8\ H_2O \rightarrow 4CH_3COOH + CO_2 + 3CH_4 + 6H_2O$

($\Delta G_0' = -26.5$ kJ/mol propionate)

The oxidation of propionate may produce methane, carbon dioxide, and acetate. Some part of propionate oxidation needs hydrogen ions. This could explain why propionate degradation was faster in a relatively acidic environment. A major part of the carbon of propionate oxidation contributed in increasing the concentration of acetate. However in ternary culture with Methnosarcina or Methanosaeta as syntrophic partners this was almost undetected due to its coupling to high rate of acetate oxidation.

The oxidation of butyrate has been studied with *Syntrophomonas wolfei* (Wofford et al., 1986) and *Clostridium bryantii* (Schink, 1991). The overall reaction of butyrate oxidation at low loading is presented in equation 50:

Equ. 50: $2CH_3CH_2CH_2COOH + 4H_2O \rightarrow 5CH_4 + 3CO_2 + 2\ H_2O$

(ΔG_0'= - 88.5 kJ/mol propionate)

In the presence of CO_2 at high loading butyrate is oxidized through (equation 51):

Equ. 51: $2CH_3CH_2CH_2COOH + CO_2 + 4H_2O \rightarrow 4CH_3COOH + CH_4 + 2H_2O$

(ΔG_0'= - 17.5 kJ/mol propionate)

By these two reactions, the oxidation of butyrate gives rise to acetate production. This explains that the increasing acetate concentration in during the degradation of VFAs might come from butyrate and also from propionate oxidation. Since oxidation of butyrate does not produce propionate, there was no contribution of butyrate oxidation in the increasing concentration of propionate.

In all batch assays, usually the concentration of acetate increased first at the beginning of the incubation time. This indicated that also other activities than the activities of propionate and butyrate converting syntrophic bacteria were involved in this process. According to Fischer (1932, cited by Dickert, 1991), upon anaerobic incubation of an anaerobic enrichment culture in the presence of either carbon monoxide or H_2 plus CO_2, acetate was formed in addition to methane. This statement clarified the phenomenon of acetate increment in anaerobic batch assay with no residual carbohydrates. Since the inoculums and the medium came from the whey reactor, carbon dioxide and sometimes H_2 gas were detected in addition to methane.

The acetate-cleaving methanogens have a very important role in the removal of acetate and by acetate cleavage they pull the butyrate oxidation reaction (Schink, 1991). This clarified that the accumulation of fatty acids started with acetate and then was followed by butyrate and propionate accumulation. When the activity of acetate-cleaving methanogens was inhibited, butyrate oxidation would be retarded. Therefore, a growth repression of one group could influence the activity of the other.

The environmental conditions influenced the conversion process of VFAs. Acetate and n-Butyrate were converted faster at neutral or slightly alkaline conditions, while propionate was converted fastest at acidic condition.

Accumulation of VFAs happened when the AnFB was supplied with whey solution at a concentration of 50 g whey$\cdot l^{-1}$ and at an organic loading rate of $\geq$ 10 kg$\cdot m^{-3}\cdot d^{-1}$. A fresh influent provided new substrate for the fermentative bacteria, which supplied fatty acids for the acidogenic bacteria. The difference in growth rate of acidogenic compared to acetogenic and methanogenic bacteria led to the observed accumulation of VFAs. The accumulation of VFAs caused a pH drop and at the end ceased the production of biogas.

A short HRT, which meant a high organic loading, caused an irreversible accumulation of VFAs (e.g. at a HRT of 4 days). This was due to the higher rates of fatty acid production than the rates of degradation during continuous feeding. A longer HRT (e.g. 6 days) apparently caused a reversible accumulation of fatty acids, since in this situation; the organic loading was less than at a HRT of 4 days. Even though fatty acids accumulated initially at a HRT of 6 days, the methanogenic bacteria were still capable to use these fatty acids after they adapted to this condition. Normally, the population in the whey reactor could adapt in about 5 days to a HRT of 6 days, when this was adjusted in one step, starting with a HRT of 12 days.

For a rough estimation of kinetic parameters of anaerobic processes Kroiss and Svardal (1999) suggested that the minimum HRT of anaerobic processes were determined by: (a) time required for acidogenesis of carbohydrates in the range of 1-22 hours, (b) time required for acetogenesis and methanogenesis of acetate in the range 1,5-13 days and (c) the time required for methanogenesis of H_2/CO_2 about or longer than 6 hours.

4.2 Kinetic Parameters of Methane Production

The kinetic parameters of methane production gave information about the activities of acidogenic, acetogenic and methanogenic bacteria in the whey reactor. The kinetic parameters were determined from the microorganism that grew in a fully suspended system. Table 18 presents the kinetic parameters of methane production with different precursor substrates for methanogenesis.

Table 18 Kinetics parameters of methane production with different precursor substrate.

Substrate	V_{max} ($mmol \cdot l^{-1} \cdot h^{-1}$)	K_m ($mmol \cdot l^{-1}$)	Conc. Substrate Max
H_2/CO_2	2.195	0.26	3 bar
HCOONa	0.883	15	200 mM
CH_3OH	0.600	30	400 mM
CH_3COONa			
- with reducing agent only	0.057	13	150 mM
- with reducing agent, trace minerals and vitamins	0.065	27	250 mM

The maximal methane production rates (V_{max}) was obtained from the experiment with an initial concentrations of the different substrates below the maximal concentration of these substrates before overloading. Therefore the specific methane production rate (v) was a function of the initial concentrations of the precursor substrate. Using the value of V_{max} and K_m, the theoretical curve of Michaelis-Menten was determined for each substrate. It was apparent that the equation of Michaelis-Menten could represent the relationship between the specific methane production rate of the whey-inoculums and the initial precursor substrate concentration of the medium.

The methane yield constant (Y) was the theoretical ratio of methane production per utilized acetate. The results showed that the "yield constant" was not constant when the substrate was the growth limiting factor. Its value depended at least on minimal initial concentrations of precursor substrate, in this case acetate. Because of this discrepancy, the term of "yield coefficient" (Y') will henceforth denote this factor. The "yield coefficient" was calculated for three different acetate concentration at optimal methanogenic condition in the presence of a reducing agent and plus or minus trace minerals and vitamins. The "yield coefficients " for acetate as a methane precursor substrate are listed in Table 19.

Table 19 Yield coefficients of methane production with acetate as the precursor substrate for methanogenesis.

Variation	Conc. (mM)	Y'_{CH4} (mmol CH_4/mmol utilized Acetate)
- with reducing agent only	10	2.053
	50	1.366
	100	0.803
- with reducing agent, trace mineral and vitamin	10	1.433
	50	0.723
	100	0.814

The "yield coefficient" (Y') at different initial acetate concentration revealed that Y' varied inversely with the initial acetate concentration. The addition of trace minerals and vitamins reduced the yield coefficient. Trace minerals and vitamins were however shortening the lag phase and enhanced and earlier methane production was seen in the assay.

4.3 Effect of Minerals Addition

Metal deficiencies can limit the performance of anaerobic digestion systems. Metal supplementation may substantially improve the performance of anaerobic digestion

(Gonzalez-Gil et al., 1999). The most important metal ions in many reactions of anaerobic digestion were Ni, Co and Fe.

In general minerals addition had a positive effect on the "yield coefficients". Ferric ions increased the yield coefficient of acetate by 50% and improved the methane and acetate conversion rate by a factor of 1.5. A combination of Ni, Co, Mn and Mg also increased the "yield coefficient" by 25% but this combination gave no significant improvement in methane and acetate conversion. The experiments using inocula from whey effluent revealed that Fe was an essential metal; for improving methanogenesis

Table 20 The effects of mineral addition on acetate and methane conversion rate and "yield coefficient".

Variation	Conversion rate ($mmol \cdot l^{-1} \cdot h^{-1}$)		Y'_{CH4}
	Acetate	Methane	(mmol CH_4/mmol utilized Acetate)
Without reducing agent (Na_2S & Resazurin)			
Control	- 0.079	+ 0.100	1.763
+ Fe^{3+}	- 0.196	+ 0.280	2.780
+ Ni^{2+}, Co^{2+}, Mn^{2+} and Mg^{2+}	- 0.063	+ 0. 110	2.135
With reducing agent (Na_2S & Resazurin)			
Control	- 0.078	+ 0.174	2.105
+ Fe^{3+}	- 0.203	+ 0.278	3.098

Iron is a required co-factor for a variety of metabolic processes in virtually all organisms (Matzanke, 1987). Iron does not only regulate the production of siderophores and the transport of protein. It is also active in regulating the formation of certain bacterial toxins and exoenzymes (Braun, 1987). A prime function of iron in aerobic microbial species is in respiration. However, it has been known for some time that anaerobic species may also require substantial quantities of iron (Neiland, 1974).

Iron is essential for microbial metabolism. However the acquisition of this element is difficult for bacteria in most natural environments where the level of free iron is typically too low to support growth. The abundance of iron in soil and sediments is essentially negated by its insoluble salts in aerobic environment, a limitation which obviously does not apply under anaerobiosis (Neilands, 1987). Ferric ions at pH 7 are not transported as metal ions but must be attached to by carrier molecules, called siderophores (Braun, 1987). The iron supply of microorganisms is intimately related to the ability of microbes to syntheses iron transport compounds, collectively designated as siderophores (Neiland, 1974).

Anaerobes and few microaerophilic growing species e.g. *Lactobacilli*, *Legionella*, *Neisseria*, fungi and *Saccaromyces cerevisiae*, appeared not to form a siderophores (Neilands, 1987). Lactic acid bacteria grew at a maximum rate in the iron-depleted medium. Lactic acid bacteria can grow anaerobically, so if iron is required at all, it would be an extremely small amount (Neiland, 1974). Enzymes from lactobacilli lack the iron containing B_2 components and have an absolute requirement of deoxyadenosylcobalamin (co-enzyme B_{12}). A variety of anaerobic bacteria in addition to *C. pasteurianum* contain flavodoxins that is a group of flavoproteins formed in the absence of iron and which have properties similar to those of ferredoxin (Neiland, 1974). The effects of iron-deficient medium on the activity of microorganism are summarized in Table 21.

Ferric ions apparently enhanced methanogenesis in batch assays and also the degradation of propionate and butyrate. The oxidation of acetate proceeded rapidly so a raise in the concentration of acetate did not appear. The total amount of acetate in the assay with ferric ions was less than in the assay without ferric ions. It seemed that ferric ions inhibited the acetogenesis from intermediate substances. This is in agreement with Kauppinen (1963) who stated that either glucose was not oxidized as completely in iron deficient cells or the yeast was unable to utilize the energy released by glucose oxidation. He mentioned also that anaerobic glucose fermentation of broken cell suspensions was found to increase with decreasing iron concentrations.

A high concentration of ferric ions in the medium had negative effects on the fermentation of glucose and lactate, since this processes involved many enzymes such as, NADH dehydrogenase and alcohol dehydrogenase. An inhibition of glucose and lactate fermentation which were the main substrates contained in whey might possibly inhibit the further process such as acidogenesis, acetogenesis and methanogenesis. This explained the negative phenomenon that was found in the implementation of ferric ions addition at a concentration of 50 $mg \cdot l^{-1}$ in the whey reactor.

Another effect of ferric ions could be the inhibition of acetogenesis. In batch assays, the concentration of acetate with ferric ions was less than in the assays without ferric iron, while in a continuous system the acetate production was varied inversely with the concentration of ferric ions addition. The mechanism of ferric ions inhibition of acetogenesis and anaerobic fermentation of glucose are still unclear. A more comprehensive study of the effect of high concentration of ferric ions on anaerobic fermentation is needed to work out in more detail the inhibition mechanism of ferric ions.

During operation of a continuous fixed-bed reactor less metals may be required than the amount required in batch suspension systems, because more biological ligands, e. g. EPS (= extracellular polymeric substances) may be produced and because continuous addition of nutrients ensures free metal availability via EPS for biomass uptake (Gonzalez-Gil et al., 1999). An excessive addition of metals as nutrients in order to enhance the treatment of wastewater should however be avoided, since this could lead to inhibitory effects on the biomass and/or to metal accumulation in the sludge. A further study to understand the processes of precipitation and dissolution of metals in anaerobic systems in a combination with mineral uptake by microorganism is necessary.

Table 21 Effects of iron-deficient medium in the activity of microorganism (Light, 1974).

Effects	Source
A. On enzymes: 1. Heme Protein Cytocrome, perioxidase, catalase concentration diminished 2. Nonheme Protein	Light, 1974
a. NADH dehydrogenase : reduced	Light, 1974
b.Succinate dehydrogenase : progressively disminished	Light, 1974
c. Ferredoxin: was not synthesized	Light, 1974
3. Other enzymes	
a, Aconitase: inhibited	Kauppinen, 1963
b.Adolase: increased (increased rate of glucose fermentation)	Kauppinen, 1963
c. Alcohol dehydogenase: reduced at a concentration of 10 mg $Fe \cdot l^{-1}$	Kauppinen, 1963
B. On morphology	
- iron deficiency: increase the length of bacteria	Winder&O'Hara, 1961, 1962 (Light, 1974)
- leads to abnormal filamentous form	Bard&Gunsalus, 1950 (Light, 1974)
- causes morphological change	(Davison, et al.,1972 (Light, 1974)
C. On growth rate	
In batch system : Specific growth rate is a fuction of the intial iron concentration in *C. guilliermodii*	Kauppinen, 1963
In continuous system : changes in the concentration of iron in the medium reservoir and therefore in the inflowing medium, can not affect the growth rate in any way.	Light, 1974

4.4 Population in Whey Reactor

Results from the batch assays demonstrated that the whey inoculums contained acetogenic bacteria, hydrogen-utilizing methanogenic bacteria, acetoclastic metanogenic bacteria, and other methane bacteria which expressed formate dehydrogenase for formate utilization. This hypothesis was based on the capability of the inocula to use H_2/CO_2, formate, methanol and acetate as substrates for methanogenesis.

The group of hydrogen-utilizing methanogens produced methane gas more rapidly than methanol and acetate-utilizing methanogens. This group had the highest maximal methane production rate. The slowest group in producing methane was the acetoclastic methane bacteria. The maximal methane production rate of this group was only 2% that of the hydrogen-utilizing methanogens. The V_{max} of this group was slightly increased by the addition of ferric ion.

The production of methane from formate was relatively slower than methane production by the hydrogen-utilizing methanogens. Formate should be cleaved to hydrogen and carbon dioxide first before it could be used for methane production. A formate dehydrogenase activity must precede methanogenesis. This process might be carried out by hydrogen-utilizing methanogens and sulfate reducers. Most of these two groups can oxidize formate (Schink, 1991). In bacterial flocs growing with whey constituents (Thiele and Zeikus, 1988), it was unclear whether hydrogen or formate was the more important electron carrier. The maximal methane production rate on formate was decreased by a factor of three as compared to the maximal methane production rate from hydrogen and carbon dioxide.

Methanogenesis from methanol can be catalyzed by metyltrophs, acetogens and hydrogenophilic methanogens (Gonzales-Gil et. al, 1999). The maximal methane production rate of this group was apparently low, about 20% of V_{max} of hydrogen-utilizing methanogens but it was still above the acetoclastic bacteria.

Methanol could be used by the acetoclastic methanogenic bacteria and some other species. The maximal production rate of this metabolic group was apparently low but it was still above of acetate cleavage by the acetoclastic methane bacteria that was 20% of V_{max} of the hydrogen-utilizing methanogens.

4.5 Comparison of Anaerobic Reactors

There are several types of anaerobic reactors, e.g. CSTR, upflow anaerobic sludge blanket (UASB) and upflow anaerobic filter (UAF). A comparison of these reactor types is listed in Table 22. These reactors are suitable for high rate anaerobic processes. A high rate anaerobic process is a process which based on highly active biomass and on biomass retention in the reactor.

Table 22 Comparison of CSTR, UASB, UAF anaerobic reactors.

Parameter	**CSTR**	**UASB**	**UAF**
Biomass retention mode	Combining mixed reactor with sludge separation and return system (Lettinga, 1983)	Formation of high settleable sludge aggregates with diameter: 0.5-2.5 mm (Lettinga, 1983; McInerney, 1999)	Bacterial attachment to particulate carrier material or immobile support structures (Lettinga, 1983)
Operation Temperature	Higher than 25°C (Lettinga, 1984)	30-40 °C (McInerney, 1999)	30-40°C (McInerney, 1999)
HRT	30-60 days (McInerney, 1999) 15-30 days at 35°C (Kroiss&Svardal, 1999)	Short, a few hours; OLR $\leq$ 40 kg·m^{-3}·d^{-3} (Lettinga, 1980)	1-10 days; OLR 4-16 kg·m^{-3}·d^{-3} (McInerney, 1999)
Wastewater	High strength waste (> 2 kg BOD·m^{-3}) (McInerney, 1999)	High and low strength wastewater	5-50 kg COD·m^{-3} (Jördening&Buchholz, 1999)
Common Problem	Liquid-solid seperation (McInernery, 1999)	Granulation process (Lettinga, 1984)	Clogging (Lettinga, 1984)
Start-up Period	Depends on indigenous population	Faster than AF (Lettinga, 1984)	1-3 months (Jördening&Buchholz, 1999)
Suitable Application	Suitable for dispersed organic matter (Grady&Lim, 1980)	Low in SS removal, fairly efficient in removing and stabilizing dispersed organic matter (Lettinga, 1984)	Good in removing and stabilizing high concentration of dispersed organic matter (Lettinga, 1984)
Process Stability	Depends strongly on the extent of acidification of the organic pollutant (Kroiss&Svardal,1999)	Depends strongly on the extent of acidification of the organic pollutant (Lettinga, 1984)	Depends strongly on the extent of acidification of the organic pollutant (Lettinga, 1984)

From the three reactor types, an AnFB reactor was the right choice for degrading whey-containing wastewater, since this wastewater has a high soluble organic matter, and only little settleable solids. This reactor can be used for degrading high strength wastewater with a relatively short HRT. However, a long of start up period and clogging by biofilm formation after a restrict operation are the problem that are commonly faced in the application of this reactor. The start up period can be reduced by improving the attachment process of the biofilm, by (1) pre-conditioning of solid surfaces, through both improving the environmental conditions and the surface itself, (2) addition of cationic polymers or slime-producing bacteria and (3) supplementing calcium (Jördening and Buchholz, 1999)

4.6 Immobilized Cell System in Anaerobic Process

More than 80% of the bacteria in different ecology have a tendency to stick to surfaces. A surface give a possibility for an accumulation of nutrients which adsorb to surfaces. This draws the microbes to adsorb to the surface and they form a biofilm or flocs.

Under condition of nutrient limitation, attachment on a surface gives some advantages for microbes, especially for anaerobic bacteria. Except for enhancing the metabolic activity, this will protect the microbes from oxygen or other stress factors like an acid pH. Surfaces may also contribute significantly to the establishment and maintenance of stable anaerobic microbial communities which depend on the benefits of any kind of interspecies metabolite transfer (Schink, 1988).

The surface film commonly known as biofilm is a term for the mode of bacterial growth on a surface of nonbiological materials. The other mode of "attached growth" is floc or granule where the new microorganisms grow on a "biological surface". Both, biofilm and flocs formation are favorable for anaerobic processes for stabilizing wastewater. Biofilms become the basis of AF (= fixed bed) reactors and flocs or granules become the basis of UASB reactor.

The primary adhesion of cells to the surface is due to hydrogen bonds, van der Wall forces and/or electrostatic interaction (Daniels, 1980). An irreversible adhesion can be reached by the production of exopolymeric substances (EPS) (Flemming, 1991).

A complex biofilm structure protects the individual cells from many adverse environmental effects. Recent studies have demonstrated that biofilms in fixed-bed reactors were highly resistant to substrate shock loads or to pH fluctuation (Schink, 1988). This becomes the major advantage of a fixed-bed reactor for wastewater treatment.

Methanothrix soehngenii, the predominant acetate degrader forms a dense network in flocs or biofilm which may cooperate in syntrophic propionate degradation.

4.7 Anaerobic Treatment of Dairy Industry Wastewater

In this work, whey was degraded in an AnFB. There were some anaerobic stationary bed system that had been applied for dairy wastewater already (Table 23).

Comparing the results full-scale reactors with the results of this work showed that at optimal conditions, a HRT in the range of 4-8 days or a OLR less than 10 kg COD·m^{-3}·d^{-1} allows around 85% removal of organic material. This demonstrated that an AnFB was a suitable alternative reactor type for treating dairy wastewater.

Table 23 Data of industrial-scale anaerobic stationary bed systems for dairy wastewater (Jördening and Buchholz, 1999).

Reactor Size (m^{-3})	Support Materrial	Load ($kg.m^{-3}.d^{-1}$)	Removal (%)	Reference
260	Plastic ring (Biofar®)	10	n.a	Weiland et al. (1988)
362	Plastic (Flocor® and cloisonyle)	10-12	70-80	Austermann-Haun et al. (1993)

In previous studies with whey, Switzenbaum and Danskin (1982) used an attached-film, expanded-bed reactor for anaerobic stabilization and biogas production. With diluted whey, a COD-removal of 90% was obtained for a loading of 14 kg $COD \cdot m^{-3} \cdot d^{-1}$. With the same loading Wildenauer and Winter (1985) could get a COD-removal efficiency of 95% with a single-stage, fixed-film loop reactor. Similar results were obtained by Boening and Larsen (1982) in a fluidized bed reactor.

Anaerobic process can be used for treating the wastewater from dairy industry. Some factor must be considered when applying the anaerobic process for this wastewater:

- It is necessary to set a pH regulator system for maintaining the pH in the range of optimal condition for methanogenesis.
- Supplementation of minerals can stimulate the anaerobic digestion significantly. Schink (1988) had mentioned the positive effect of nickel and iron as trace element for methanogenic bacteria. The minerals must not be added excessively in order to prevent an inhibition effect of this addition, if the mineral concentration is getting to high.
- A fixed bed reactor could be an alternative of reactor type to a continuously stirred tank reactor since for anaerobes, attached growth can enhance metabolic activity of the adsorbed microbial population. A biofilm also has the advantage of an efficient protection from oxygen.

Some general requirements of materials as a support for bacterial biofilms in fixed-bed reactor were summarized by Jördening and Buchholz (1999):

- Availability of material in big quantity (> 100 m^3).
- Low costs of the material
- Inert behavior (mechanically and microbially) without toxic effects and easy disposal.

- Low pressure drop (low energy demand for mixing of fluidization).

In 1995, 26.74 million tons of milk were delivered to the dairies of the Federal Republic of Germany. This makes an annual consumption of approximately 330 litre of milk per capita (Rosenwinkel et. al. 1999). The wastewater in dairy industry mostly comes from the cleaning process of transportation equipment and the production plant. This contains more than 90% of the pollutants from milk. The average amount of wastewater of dairy is 1-2 m^3 for 1000 kg milk that was processed to products not including the wastewater from cooling and evaporating processes. The concentration of COD of dairy wastewater is in the range of 350-1400 mg $COD \cdot l^{-1}$. The pH value is 9-10.5 due to the caustic solution for bottle cleaning in the washing machines (Rüffer and Rosenwinkel. 1991). The wastewater from the cheese industry has a higher concentration of organic content, contains NaCl and has a relatively low pH. Dairy wastewater has only a small proportion of settleable solids.

The major problems in handling dairy wastewater are the variations in wastewater volume and concentrations over one day production. The minimum requirements of wastewater treatment units in dairy industry are listed in Table 24.

Table 24 Minimum requirements of wastewater treatment units in dairy wastewater treatment plant.

Unit	Purpose
Sand Trap	Separating big particle and sand in the wastewater.
Equalizing Tank	- Mixing the different concentration of wastewater. - Equalizing the variation of wastewater flow. - Neutralizing the pH of wastewater.
Flotation (can be placed at pretreatment stage, or can replacing equalizing tank or the post-treatment stage)	Removing of fats and protein in the wastewater
Biological Treatment	Stabilizing the wastewater (degrading the organic material content of wastewater) Preferably an anaerobic process.
Clarifier	Separating the biomass which is washed out or slaughed off from the biological reactor.

The equalizing tank should be designed with a retention time around less than 1 and 1.5 days. The minimum volume is approximately 25% of the daily wastewater flow. Normally the equalizing tank does not contained an aeration system, thus pre-acidification may occurre in this tank. Lactose is mainly converted to lactate, propionate, and butyrate. When the pH in the equalizing tank is controlled correctly, it will give an advantage for the anaerobic treatment. The anaerobic process will receive a soluble wastewater mainly containing volatile fatty acids. This will avoid the accumulation of VFAs because the pH has already adjusted in the previous unit and the methane reactor is designed based on the basis of time required for acetogenesis and methanogenesis of VFAs. Without any possibility of a further accumulation of VFAs, the reactor does not require a pH regulator system.

The results of whey degradation experiments in laboratory scale, gave some information that were necessary for wastewater treatment in a dairy company. Whey can represent the characteristic composition of the wastewater from dairy industry. The experiments of this work tried to simulate the degradation of a highly concentration wastewater containing to 50 kg COD·m^{-3}. Since the results demonstrated that an AnFB had a good performance in organic removal and methane production with such a concentration of organic compounds at a HRT of 5 days, this promised that an AnFB reactor could be applied for treating dairy industry wastewater. The actual wastewater dairy plant has a lower concentration of organic material than that in the simulation experiments. When an AnFB reactor was used for dairy industry wastewater (=350-1400 mg COD·l^{-1}) the performance may be possibly better than in the simulation experiment, either in removing of organic material and in producing methane at the same HRT as in the laboratory or it could be operated at a shorter HRT (= a higher OLR).

Chapter 5

Summary

5.1 Conversion of Volatile Fatty Acids (VFAs)

The accumulated VFAs in the AnFB reactor were rapidly degraded, when the supply of fresh feeding was stopped for recovery of a highly loaded or overloaded system. VFAs conversion rates in whey at different initial pH can be seen in Table 25.

Table 25 VFAs conversion rates at different pH.

VFA	Unit	Initial pH		
		6.4	6.3	7.25
Acetate	$mmol \cdot h^{-1}$	- 0.275	- 0.536	- 1.050
Propionate	$mmol \cdot h^{-1}$	- 0.063	- 1.880	- 0.050
n-Butyrate	$mmol \cdot h^{-1}$	- 0.038	- 0.054	- 0.170

The pH in the reactor influenced the conversion process of VFAs. Acetate and n-Butyrate were converted faster at neutral or slightly alkaline conditions, while propionate was degraded faster at slightly acidic conditions.

5.2 Accumulation of VFAs and Biogas Production

Accumulation of VFAs happened when the AnFB was supplied with whey solution (50 g $whey \cdot l^{-1}$) after restart at short hydraulic retention times (= high volumetric loading). Accumulation rates of VFAs at different HRT are compared in Table 26.

VFAs were accumulated faster at a HRT change from 12 to 6 days compared to a HRT change from 6 to 4 days. However, at a HRT of 6 days, the accumulated VFAs were degraded after an adaptation period of about 5 days, whereas the accumulated VFAs at a HRT of 4 days remained constant upon time.

Table 26 Accumulation rates of VFAs in the whey reactor.

Item	Unit	HRT (d)			
		4 stop→ 4	6 stop→ 6	6 (after oxygenation)	6 (after a HRT of 12d)
Acetate	$mmol \cdot h^{-1}$	+ 0.11	+ 0.58	+ 0.43	+ 0.11
Propionate	$mmol \cdot h^{-1}$	+ 0.33	+ 0.25	+ 0.18	+ 0.02
n-Butyrate	$mmol \cdot h^{-1}$	+ 0.03	+ 0.06	+ 0.11	+ 0.03
Biogas	$l \cdot h^{-1}$	+ 0.03	+ 0.02	n.a	n.a
pH	$unit \cdot h^{-1}$	- 0.04	- 0.03	- 0.11	- 0.004

Oxygenation caused an accumulation of VFAs to a similar extent as the accumulation event at a HRT of 4 days. Shortening the HRT from 12 days to 6 days (= doubling of the loading), led to an accumulation of VFAs. The accumulated VFAs were degraded after the adaptation period (= around 5 days).

5.3 Kinetics Parameters of Methane Formation

The kinetic parameters of different precursors of methanogenesis were investigated in batch assays. The kinetic parameters were calculated using the Michelis-Menten model. The results are listed in Table 27:

Table 27 Kinetic constants for methane formation with different substrates.

Substrate	V_{max} ($mmol \cdot l^{-1} \cdot h^{-1}$)	K_m ($mmol \cdot l^{-1}$)
H_2/CO_2	2.438	24
HCOONa	0.883	755
CH_3OH	0.905	2549
CH_3COONa		
- w/o reducing agent & minerals	n.a	n.a
- with reducing agent only	0.057	13
- with reducing agent, trace minerals and vitamins	0.065	27

H_2/CO_2 was the best methanogenic substrate in the whey reactor, followed by formate and methanol. The least degradable substrate in suspension cultures was acetate. Since acetate is an obligate intermediate product of glycolysis and since it was degraded in the whey reactor quite rapidly, the acetoclastic methanogens might have been presented mainly as component of the biofilm on the carrier material.

5.4 Improvement of Acetate Degradation and Methane Production by Supplying Trace Minerals

The conversion rates of acetate and the methane production rates of batch assays are listed in Table 28:

Table 28 Acetate degradation and methane production rates.

Variation	Conversion rate ($mmol \cdot l^{-1} \cdot h^{-1}$)	
	Acetate	**Methane**
Without reducing agents (Na_2S & Resazurin)		
Control	- 0.079	+ 0.100
+ Fe^{3+}	- 0.196	+ 0.280
+ Ni^{2+}, Co^{2+}, Mn^{2+} and Mg^{2+}	- 0.063	+ 0. 110
With reducing agents (Na_2S & Resazurin)		
Control	- 0.078	+ 0.174
+ Fe^{3+}	- 0.203	+ 0.278

Ferric ions addition or the addition of a mix of minerals improved acetate degradation and methane production rates more than two-folds. The redox potential $\pm$ addition of a reducing agent was low enough for methanogenesis.

5.5 Population

The suspended part of the population inside the AnFB reactor was incapable in converting acetate to methane. This was due to (1) an insufficient number of acetoclastic methanogenic bacteria, in the suspended portion of methanogens and (2) unsuitable conditions inside the reactor for methanogenic bacteria, especially a pH below 7, which was caused by fatty acids accumulation at little buffer capacity.

The population was also capable in reducing accumulated propionate, if the residence time was longer or at least equivalent to the growth rates. The H_2, which was produces from (a) glycolysis of lactose or from lactate and (b) from acetogenesis of propionate and butyrate could be converted by the population. This indicated that there were significant numbers of active H_2/CO_2-utilizing methanogenic bacteria.

The population inside the reactor contained acetogenic and methanogenic bacteria, but these bacteria grew slower than acidogenic bacteria. The metabolic activities of acetogenic and methanogenic bacteria were slow compared to that of acidogenic bacteria. This made the acid degradation rate < the acid production rate. The minimal HRT was thus dependent on acid degradation rates. The population was incapable in preventing the VFAs accumulation when the pH in the reactor decreased below the optimal pH for the methanogenic population.

The population in the AnFB contained hydrogen-utilizing methanogenic bacteria, formate-utilizing methanogenic bacteria, methanol-utilizing methanogenic bacteria, acetoclastic methanogenic bacteria and sulfate-reducing bacteria.

The activity of methanogenic bacteria ceased almost completely after the population was contacted with oxygen (oxygenation). Acidogenic bacteria had a much higher oxygen tolerance compared to methanogenic bacteria.

Acetate-utilizing methanogens seemed to be unable to grow as in suspension. Acetoclastic methanogens preferred to grow in a particulate form within flocs attached on a support material, since flocs or the biofilm on the support material provide a lower redox potential and anaerobic conditions that were needed by these bacteria. The additional of a reducing agent was necessary for culturing acetoclastic methanogens in suspended cultures. Without the reducing agent, the methane production was only 10% of its theoretical amount.

The optimal H_2 gas concentration for methanogens was provided at 2.25 bar H_2/CO_2 (80%:20%) in gas phase above the liquid. Increasing the amount of H_2 gas beyond this value led to lower methane production rates.

5.6 Effects of pH and Feeding

In the whey reactor a decrease of pH of around 1 unit occurred when the acetate concentration increased by 22.5-25 mM. A drop of pH by one unit inside the reactor would influence and even prevent the growth of methanogenic bacteria and methanogenesis. Higher volatile fatty acid concentrations could be tolerated if a pH regulator system was used for keeping the pH inside the reactor in an optimal range for methanogenic bacteria.

A pH control at $\pm$ 7 could prevent a complete breakdown of methanogenesis during accumulation of VFAs at high OLR. When treating whey, the system should be equipped with pH regulation since whey itself has a low buffering capacity. A pH regulator was absolutely essential for OLR's of $\geq$ 10 kg COD$\cdot$ $m^{-3}\cdot d^{-1}$ in the whey reactor.

5.7 Effects of Mineral Addition

In batch fermentation, reducing agents (Na_2S and Resazurin) was required for culturing acetoclastic methanogens in suspended cultures. The ability of acetoclastic methanogens in converting acetate to methane was improved by trace mineral and vitamin additional.

Ferric ions at concentration of 50 $mg\cdot l^{-1}$ improved the methanogenesis and acetate degradation. Nickel (II) and cobalt (II) at a final concentration of 50 $mg\cdot l^{-1}$ for each ion had an inhibitory effect on methanogenic bacteria, either as single ions or in a combination of both ions. Lower concentrations of Ni and Co were essential ions for methanogenesis. Manganese (II) could reduce the inhibition effect of Ni and Co.

Chapter 6

References

Ahring, B. K. 2003. Perspectives for Anaerobic Digestion. In: *Biomethanation I. Advances in Biochem. Eng. / Biotechnol.* No. 81, pg. 1–30. Scheper, T.(series ed.). Ahring, B. K.(vol. ed). Springer Verlag. Berlin, Heidelberg, New York.

Angelidaki, I., L. Ellegaard and B. K. Ahring. 2003. Application of Anaerobic Digestion Process. In: *Biomethanation II. Advances in Biochem. Eng./Biotechnol.* No. 82, pg. 1-34. Scheper, T.(series ed.). Ahring, B. K.(vol. ed). Springer Verlag. Berlin, Heidelberg, New York.

APHA. 1989. *Standard Methods for the Examination of Water and Wastewater.* 16th edition. American Public Health Association. Washington, D.C.

Benefield, L. D. and C.W. Randal. 1980. *Biological Process Design for Wastewater Treatment.* Prentice-Hall, Inc. Englewood Cliffs.

Boening, P. H. and V. F. Larsen. 1982. Anaerobic fluidized bed whey treatment. *Biotech. Bioeng.* 24: 2539-2556.

Böhnke, B., W. Bischofsberger and C. F. Seyfried. 1991. *Anaerobtechnik Handbuch der anaeroben Behandlung von Abwasser und Schlamm.* Springer Verlag. Berlin, Heidelberg, New York.

Braun, V., K. Hantke, K. Eick-Helmerich, W. Köster, W. Preßler, M. Sauer and S. Schöffer. 1987. Iron transport system in Eschericia coli. In: Iron Transport in Microbes, Plants and Animals. Winkelmann, G., D.van der Helm and J. B. Neilands (eds.). VCH. Weinheim.

Conway de Macario, E. and A. J. K. Macario. 2003. Molecular Biology of Stress Gene in Methanogens: Potential for Bioreactor Technology. In: *Biomethanation I, Advances in*

Biochem. Eng./Biotechnol. No. 81. pg. 95-150. Scheper, T.(series ed.). Ahring, B. K.(vol. ed). Springer Verlag. Berlin, Heidelberg, New York.

Daniels, S. L. 1980. Mechanisms involved in sorption of microorganisms to solid surfaces In: *Adsorption of Microorganisms to Surfaces*. Bitton, G. and K. C. Marshall (eds.). John Wiley & Sons. New York. pg. 7-58.

Diekert, G. 1991.Acetogenic Bacteria. In: *The Prokaryotes*. 2nd edition. Balows, A., H. G. Trüper, M. D.Dworkin, W. Harder and K-H. Schleifer (eds.). Volume I. Springer-Verlag. New York, Berlin, Heidelberg.

Elsden, S. R. 1952. *The enzymes*. Summer, J. B. and K. Myrbäck (ed). Vol. 2. pg. 791. Academic Press., New York.

Flemming, H. -C. 1991. Biofilme und Wassertechnologie, Teil 1: Entstehung, Aufbau und Zusammensetzung. *gwf Wasser Abwasser* 132: 197-207.

Gallert, C. and J. Winter. 1997. Mesophilic and thermophilic anaerobic digestion of source-sorted organic waste: effect of ammonia on glucose degradation and methane production. *App. Microbiol. Biotechnol.* 48: 405-410.

Gavala, H. N., I. Angelidaki and B. K. Ahring. 2003. Kinetics and Modeling of Anaerobic Digestion Process. In: *Biomethanation I, Advances in Biochem. Eng. Biotechnol.* No. 81. pg. 57-94. Scheper, T.(series ed.). Ahring, B. K.(vol. eds.). Springer Verlag. Berlin, Heidelberg, New York.

Gillies, M. T. 1974. *Whey Processing and Utilization: Economic and Technical Aspect.* Noyes Data Corp. New Jersey.

Gonzales-Gil, G., R. Kleerebezem, and G. Lettinga. 1999. Effects of nickel and cobalt on kinetics of methanol conversion by methanogenic sludge as assessed by on-line CH_4 monitoring. App. Environ. Microbiol. 65:1789-1793.

Gottschalk, G. 1979. *Bacterial Metabolism*. Springer-Verlag. New York, Heidelberg, Berlin

Grady Jr, C. P. L. and H.C. Lim. 1980. *Biological Wastewater Treatment: Theory and Applications.* Marcel Dekker, Inc.. New York and Basel.

Hartmann, L. 1992. *Biologische Abwasserreinigung.* 3rd edition. Springer-Verlag. Berlin, Heidelberg, New York.

Hobson, P. H. and R. Summers. 1967. The Continuous Culture of Anaerobic Bacteria, *Journal of General Microbiology.* 47. pg. 53.

Jördening, H. J. and K. Buchholz. 1999. Fix-film stationary and fluidised bed reactor. In: Biotechnology. Rehm, H.-J and G. Reed (eds.). Vol. 11a. Wiley-VCH. Weinheim, New York, Chichster. pg. 493-516

Kauppinen, V. 1963. *The effect of iron deficiency on the growth and glucose metabolism of Candida guilliermondii.* Annales Academiae Scientiarum Fennicae. Series A, II Chemica. 123. Helsinki.

Kosaric, N. and Y. J. Asher. 1985. The Utilization of Cheese Whey and Its Component. In: *Agricultural Feedstock and Waste Treatment and Engineering.* Springer-Verlag, Berlin.

Kosikowski, F. V. 1982. *Cheese and Fermented Milk Food.* 2nd ed. Kosikowski, F. V. and Associates. Brooktondale, New York. pg. 711.

Kroiss, H and K. Svardal. 1999. CSRT-reactors and contact process in industrial wastewater treatment. In: *Biotechnology.* Rehm, H.-J and G. Reed (eds.). Vol. 11a. Wiley-VCH. Weinheim, New York, Chichster. pg. 479-492.

Kugelman, Irwin J. and K. K. Chin. 1971. *Toxicity, Synergism, and Antagonism in Anaerobic Waste Treatment Process in Anaerobic Biological Treatment Process.* American Chemical Society. Washington D. C. pg. 55–90.

Lettinga, G. 1984. The prospects of anaerobic wastewater treatment. In: *Anaerobic Digestion and Carbohydrate Hydrolysis of Waste*. Ferro, G. L., M. P. Ferranti and H. Caveau (eds.) Elsevier App. Sci. Pub. London, New York. pg. 262-273.

Lettinga, G. A., F. M. van Velsen, S. W. Homba, W. Zeeuw and A. De Klapwijk. 1980. Use of the upflow sludge blanket (USB) reactor concept for biological wastewater treatment, especially for anaerobic treatment. *Biotechnol. Bioeng.* 22: 699-734.

Lettinga, G., S. W. Hobma, L. W. Hulshöff Pol, W. De Zeeuw, P. de Jong, P. Grin and R. Roersma. 1983. Design, operation and economy of anerobic treatment. *Wat. Sci. Techn.* 15: 177-195.

Leviton, Abraham and Elmer H. Marth. 1965. Fermentation. In: *Fundamentals of Dairy Chemistry*, Webb and Johnson (ed). The Avi Publishing Company Inc. Westport-Connecticut. pg. 673-770.

Light, A. and R. A. Clegg. 1974. Metabolism in iron-limited growth. In: *Microbial Iron Metabolism, A Comprehensive Treatise*. Neilands, J. B. (eds.). Academic Press. New York, London.

Matzanke, B. F. 1987. Mössbauer spectroscopy of microbial iron uptake and metabolism In: *Iron Transport in Microbes, Plants and Animals*. Winkelmann, G., D.van der Helm and J. B. Neilands (eds.). VCH. Weinheim.

McInerney, M. 1999. Anaerobic metabolism and its regulation. In: *Biotechnology*. Rehm, H.-J and G. Reed (eds). Vol. 11a. Wiley-VCH. Weinheim, New York, Chichster. pg. 455-478.

McInerney, M. J. 1988. Anaerobic degradation of protein and lipids. In: Zehner, A J. B. (ed). *Biology of Anaerobic Microorganism*, John Wiley and Sons. New York. Pg. 373-415.

Neilands, J. B., K. Konoke, B. Schwyn, M. Coy, R. T. Fancis, H. Paw and A. Bagg. 1987. Comparative biochemistry of microbial iron assimilation. In: Iron Transport in Microbes, Plants and Animals. Winkelmann, G., D.van der Helm and J. B. Neilands (eds.). VCH. Weinheim.

Neilands, J. B.1974. Iron and its role in microbial physiology. In: Microbial Iron Metabolism, A Comprehensive Treatise. Neilands, J. B. (eds.). Academic Press. New York, London.

Prescott, S. C., and C. G. Dunn. 1949. *Industrial Microbiology.* McGraw-Hill Book Co. New York. pg. 5, 314.

Rosenwinkel, K. –H and U. Austermann-Haun, and H. Meyer. 1999. Industrial wastewater source and treatment strategies. In: Biothecnology, 2. Finn, R. K. and P. Präve (eds.). Hanser Publishers. Munich. pg. 191-216.

Rüffer, H. and K-H. Rosenwinkel. 1991. Taschenbuch der Industrieabwasserreinigung. R. Oldenbourg Verlag. München, Wien.

Sánchez, E., R. Borja, P. Weiland, L. Travieso and A. Martin. 2000. Effect of temperature and pH on the kinetics of methane production, organic nitrogen and phosphorus removal in the batch anaerobic digestion process of cattle manure. Bioprocess and Biosystems Eng. 22 (3): 0247-0252.

Schink, B. 1990. Conversion of small amounts of energy in fermenting bacteria. In: Biotechnology, 2. Finn, R. K. and P. Präve (eds.). Hanser Publishers. Munich. pg. 63-89.

Schink, B. 1991.Syntrophism among Prokaryotes. In: *The Prokaryotes.* 2nd edition. Balows, A., H. G. Trüper, M. D.Dworkin, W. Harder and K-H. Schleifer (eds.). Volume I. Springer-Verlag. New York, Berlin, Heidelberg.

Segel. I.H. 1968. *Biochemical Calculation.* John Wiley and Sons, Inc. New York.

Skiadas, I. V., H. N. Gavala, J. E. Schmidt and B. K. Ahring. 2003. Anaerobic Granular Sludge and Biofilm Reactors. In: *Biomethanation II, Advances in Biochem. Eng./Biotechnol.* No. 82, pg. 35-68. Scheper, T.(series ed.). Ahring, B. K.(vol. ed). Springer Verlag. Berlin, Heidelberg, New York.

Stams, A. J., S. J. W. H. Oude Elferink and P. Westermann. 2003. Metabolic Interaction Between Methanogenic Consortia and Anaerobic Respiring Bacteria. In: *Biomethanation I,* Advances in Biochem. Eng./Biotechnol. No. 81. pg. 31–56. Scheper, T.(series ed.). Ahring, B. K.(vol. ed). Springer Verlag. Berlin, Heidelberg, New York.

Switzenbaum, M. S. and S. C. Danskin. 1982. Anaerobic expanded bed treatment of whey. *Agr. Waste.* 4: 411-426

Thiele, J. H. and J. G. Zeikus. 1988. Control of interspicies electron flow during anerobic digestion: significance of formate transfer versus hydrogen transfer during syntrophic methanogenesis in flocs. App. Environ. Microbiol. 53: 24-29.

Ueno, Y., S. Haruta, M. Ishii and Y. Igarashi. 2001. Changes in product formation and bacterial community by dilution rate on carbohydrate fermentation by methanogenic microflora in continuous flow stirred tank reactor. App. Microbiol. Biotechnol. 57 (1-2): 65-73.

Vogels, Godfried D., Jan T. Keltjens and Chris van der Drift. 1988. Biochemistry of Methane Production. In: *Biology of Anaerobic Microorganisms*, Zehnder, Alexander J. (ed), John Wiley & Sons. New York. pg. 707–770.

Walstra. P., T. J. Geurts, A. Noomen, A. Jellema and M. A. J. S. van Boekel. 1999. *Dairy Technology: Principle of Milk Properties and Process.* Marcel Dekker, Inc. New York-Basel.

Widdel, F. 1988. Microbiology and ecology of sulfate- and sulphur-reducing bacteria. In: *Biology of Anaerobic Microorganisms*, Zehnder, A. J. (ed), John Wiley & Sons. New York. pg. 469-585.

Wildenauer, F. X and J. Winter. 1985. Anaerobic digestion of high-strength acidic whey in a pH-controlled up-flow fixed film loop reactor. *App. Microbio. Biotechnol.* 22: 367-372.

Wofford, N. Q., P. S. Beaty and M. J. McInerney. 1986. Preparation of cell-free extracts and the enzymes involved in fatty acid metabolism in Syntrophomonas wolfei. *J. Bacteriol.* 167: 179-185.

Wolf, P. and W. Nordmann, 1977. Eine Feld-Methode für die Messung des CSB von Abwässern. *Korrespondenz Abwasser* 24, No. 9: 277-281.

Wozniewski, T. 1990 *Analytische und Biochemische Untersuchungen mit pflanzlichen Glucomannanen*. Dissertation an der Universität Regensburg. FB Pharmazie, pg.121.

Zayed, G. and J. Winter. 1998. Removal of organic pollutant and of nitrate from wastewater from dairy industry by denitrification. *App. Microbiol. Biotechnol.* 48: 469-474.

Zayed, G. and J. Winter. 2000. Inhibition of methane production from whey by heavy metals – protective effect of sulfide. *App. Microbiol. Biotechnol.* 53: 726-731.

Zeikus, J. G. 1977. The biology of methanogenic bacteria. *Bacteriol. Rev.* 41: 514-541.

Zinder, S. H. 1984. Microbiology of anaerobic conversion of organic waste to methane: recent development. *ASM News* 50: 294-386.

Chapter 7

Appendix: Transparences of Oral Presentation

Oral Presentation of PhD. Thesis

Degradation of Whey in an Anaerobic Fixed Bed (AnFB) Reactor

"Stabilisierung von Süβmolke in einem Festbettreaktor mit Liapor-Blähton-Polyethylene-Sinter Lamellen"

Marisa Handajani, MT.

23.07.2004

Faculty of Civil Engineering, Geology and Surveying

Universität Karlsruhe (TH)
Institut für Ingenieurbiologie und Biotechnologie des Abwassers

Topic

- Background
  - Milk, dairy wastewater and whey production
  - Anaerobic degradation of whey
- Results:
  - Experiments in an anaerobic fixed bed reactor
    - → Observe the optimum operating parameters
  - Experiments in batch assays
    - → Observe the kinetic parameters
    - → Effect of trace elements
- Conclusion

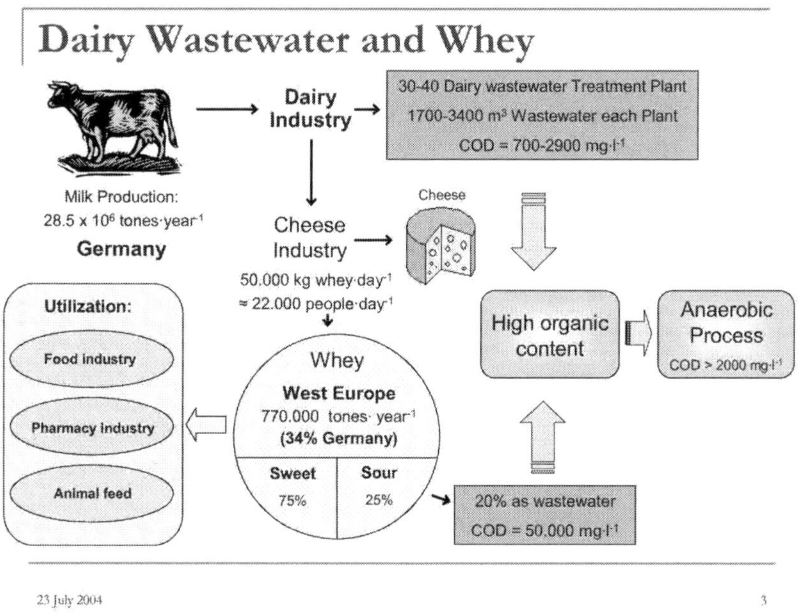
Dairy Wastewater and Whey
Dairy Industry
30-40 Dairy wastewater Treatment Plant
1700-3400 m³ Wastewater each Plant
COD = 700-2900 mg·l⁻¹
Milk Production:
28.5 x 10⁶ tones·year⁻¹
Germany
Cheese Industry
Cheese
50.000 kg whey·day⁻¹
≈ 22.000 people·day⁻¹
Utilization:
Food industry
Pharmacy industry
Animal feed
Whey
West Europe
770.000 tones· year⁻¹
(34% Germany)
Sweet
75%
Sour
25%
High organic content
Anaerobic Process
COD > 2000 mg·l⁻¹
20% as wastewater
COD = 50.000 mg·l⁻¹
23 July 2004
3

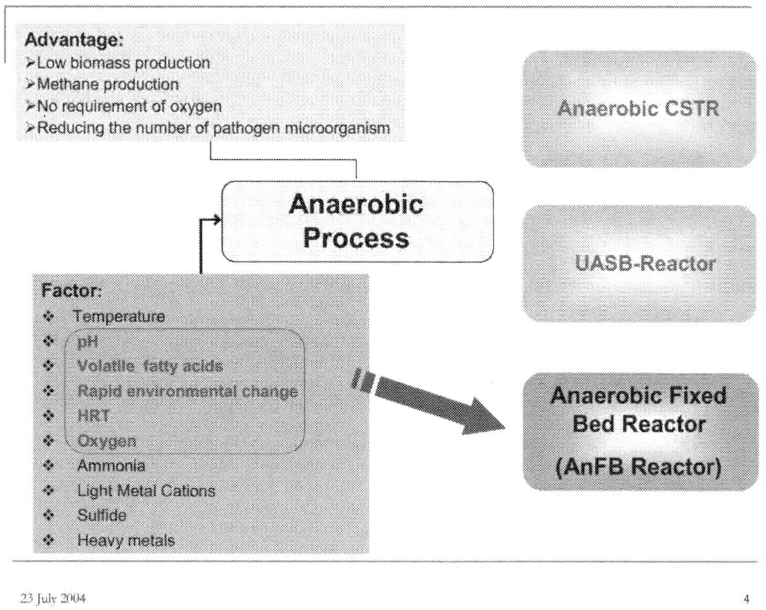
Advantage:
➢Low biomass production
➢Methane production
➢No requirement of oxygen
➢Reducing the number of pathogen microorganism
Anaerobic CSTR
UASB-Reactor
Anaerobic Process
Factor:
Temperature
pH
Volatile fatty acids
Rapid environmental change
HRT
Oxygen
Ammonia
Light Metal Cations
Sulfide
Heavy metals
Anaerobic Fixed Bed Reactor
(AnFB Reactor)
23 July 2004
4

Whey: anaerobic food chain

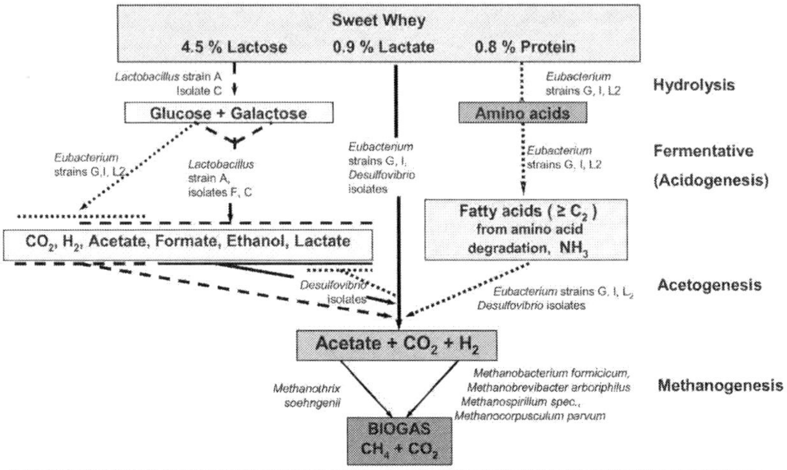

23 July 2004 5

Objective

- Determine the performance of an AnFB Reactor for treating whey-containing wastewater.
- Investigate the parameters that trigger the operational failure.
- Optimize the anaerobic degradation of whey-containing wastewater in a anaerobe fixed bed reactor.
- Determine the kinetic parameters for degradation of whey-containing wastewater.

23 July 2004 6

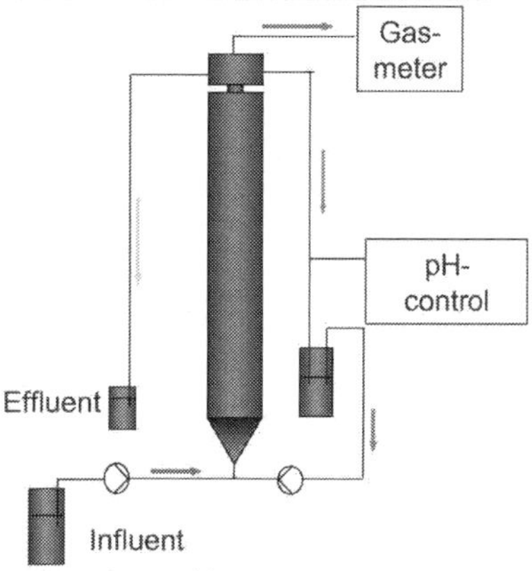

Operating condition:

Volume : 4.5 l liquid

Influent : Whey solution (50 g COD·l^{-1})

Temp. : 37°C

Scheme of AnFB Reactor

Anaerobic Fixed Bed Reactor

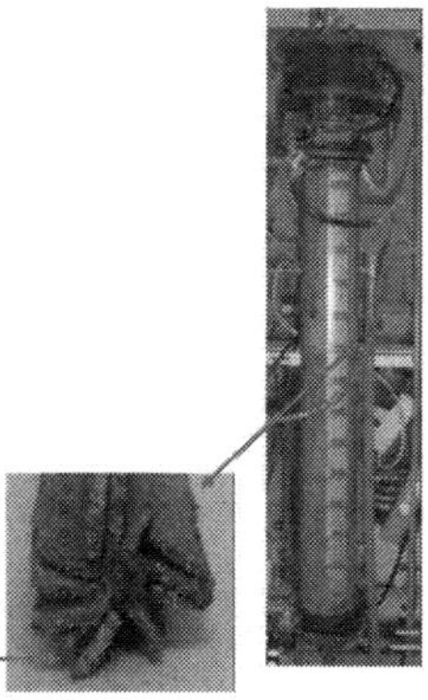

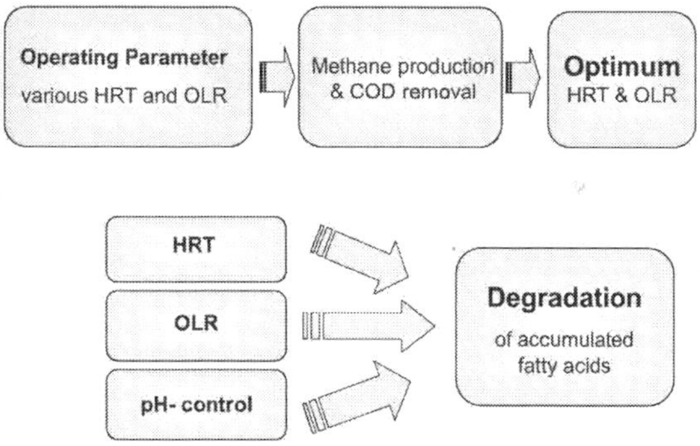

Liapor-clay-polyethylene-sinter lamella

(Herding Co., Amberg).

Size: 400 x 40 x 4 mm

AnFB Reactor

Dependency of COD removal on HRT and OLR

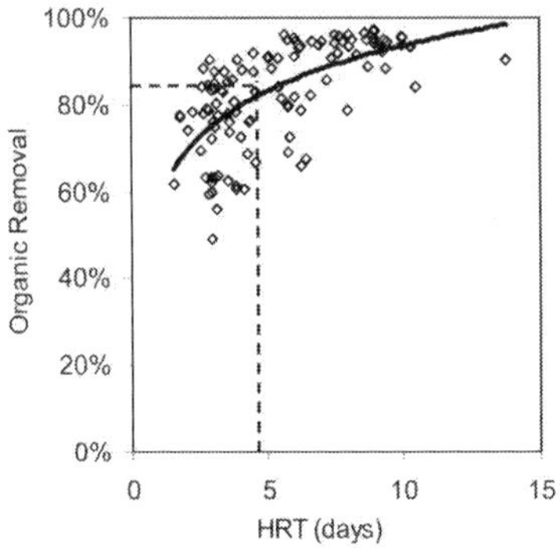

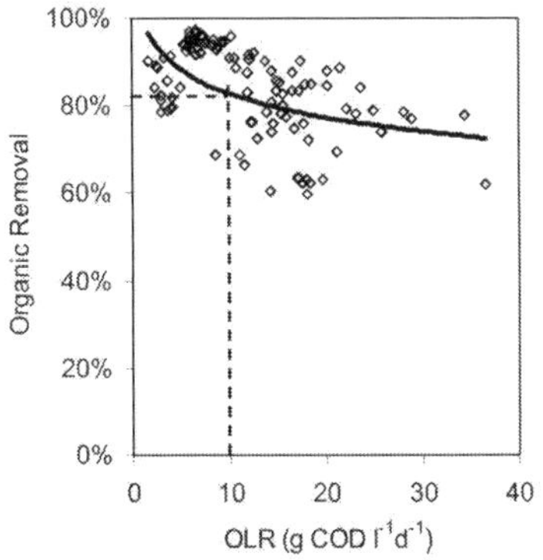

By a HRT of 5 days ≈ an OLR of 10 g COD·l^{-1}·d^{-1}, 85% of COD removal was reached.

23 July 2004 9

AnFB Reactor

Dependency of residual fatty acids in the effluent on HRT and OLR

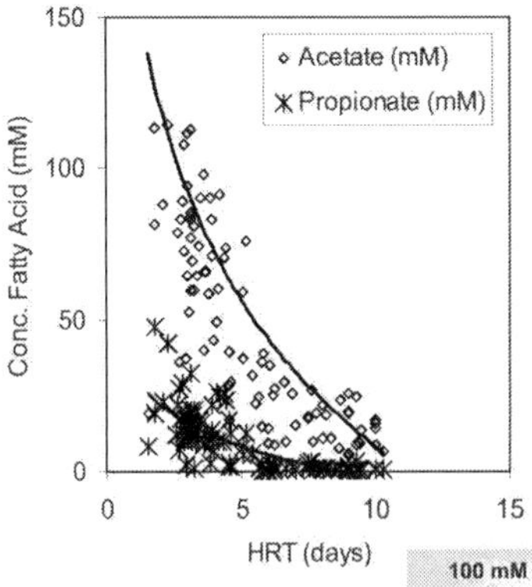

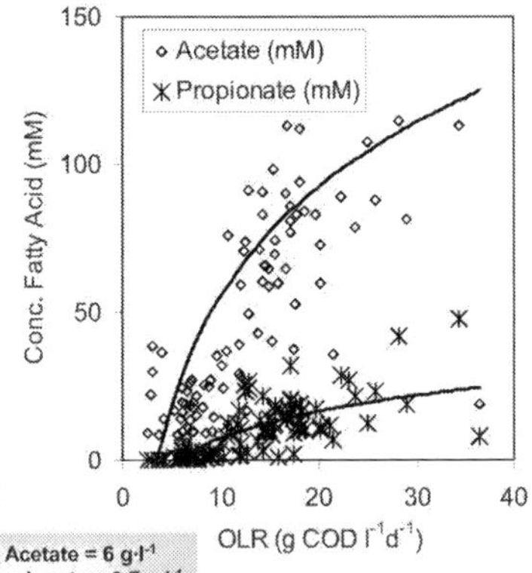

100 mM Acetate = 6 g·l^{-1}
50 mM Propionate = 3.7 g·l^{-1}

Increasing OLR (≈ shortening HRT) accumulated the acetate and propionate in the effluent of AnFB → decreased the COD removal.

23 July 2004 10

AnFB Reactor

Dependency of methane production on HRT and OLR

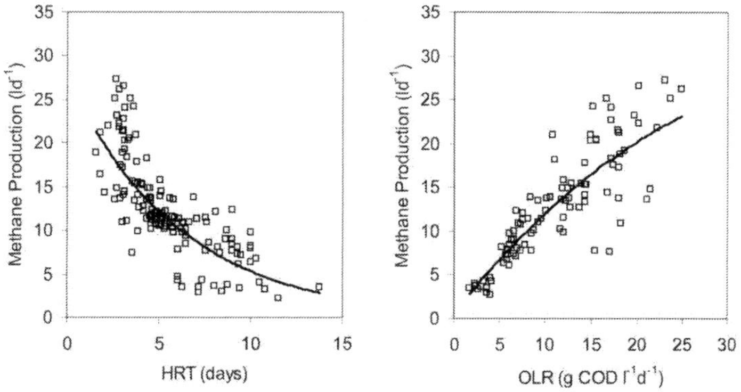

By an OLR of 10 g COD·l^{-1}·d^{-1} and a HRT of 5 days gave the specific methane production ≈ 0.30 l CH_4·g^{-1} COD removed (≈ 85% η COD)

23 July 2004 11

AnFB Reactor

pH-value and fatty acids at an equilibrium condition for a HRT of 12, 8, and 4 day

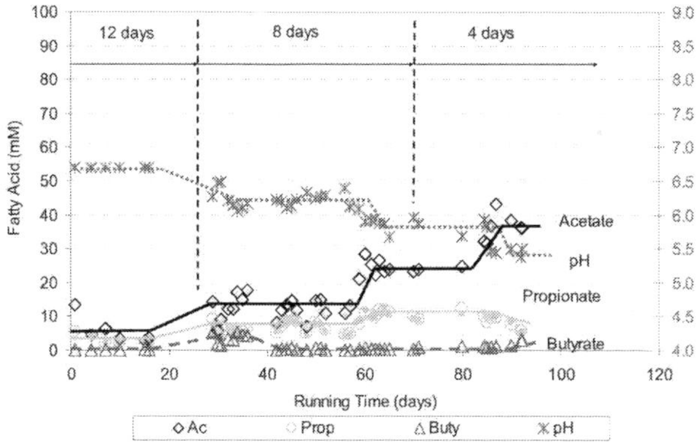

Without pH-regulation in the reactor, the pH dropped to pH < 5.5 by shortening the HRT from 12 to 4 day (≈ fatty acids accumulation)

23 July 2004 12

AnFB Reactor

Biogas production by various HRT

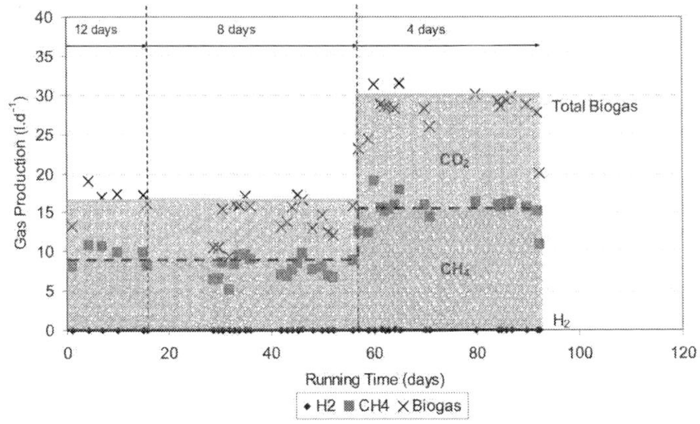

By shortening the HRT from 12 to 4 day, the CO_2 content in biogas increased (pH < 5,5); hydrogen was not detected in the biogas.

AnFB Reactor

Fatty acids concentrations during starving and feeding phases

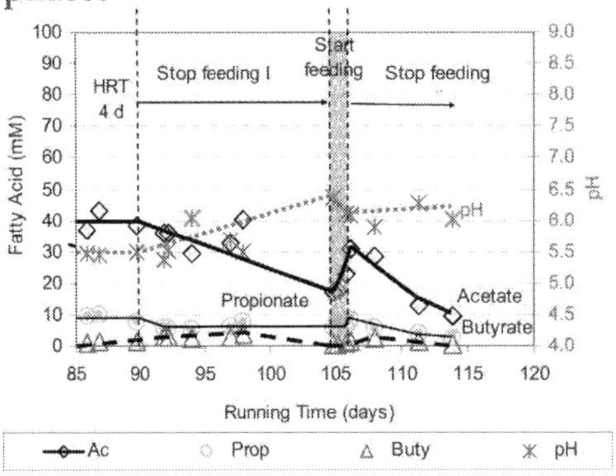

Degradation and accumulation rates

Variation	**Acetate** (mmol·hr^{-1})
First Stop (0 g whey·l^{-1})	-0.275
Feeding (50 g whey·l^{-1}; HRT 4 days)	1.125
Second Stop (0 g whey·l^{-1})	-0.536

The acetate degradation rates depended on the initial concentration and pH-value. At a HRT of 4 days, the acetate production rates was higher than acetate reduction rates → Accumulation.

AnFB Reactor

Effect of different HRT on the degradation of accumulated fatty acids after an oxygenation

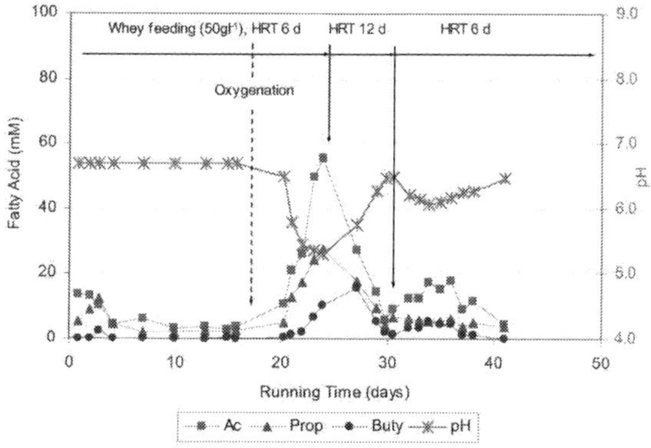

"Oxygenation Accident" led an accumulation of fatty acids → pH dropped; Longer HRT (≈ Reducing the OLR) recovered the reactor ➔There was no irreversible damage on anaerobic population.

AnFB Reactor

Fatty acids degradation with pH-regulation

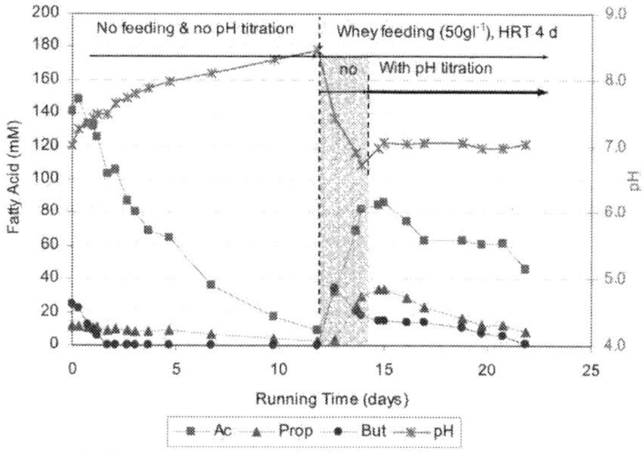

By installing the pH-value regulation in the reactor, the fatty acids could be degraded during feeding and a HRT of 4 day (≈ OLR of 12.5 g $COD \cdot l^{-1} \cdot d^{-1}$) → η COD increased

Kinetic parameters of methane production with different substrates for methanogens bacteria

CH_4-Production rate, V_{max} und K_m

Substrate:
- H_2/CO_2
- Na-Formate
- Methanol
- Na-Acetate

Condition:
- Suspended microorganism in whey-reactor effluent 20% (vol/vol)
- In orbital shaker at 37°C
- Under anaerobe condition(N_2 Gas atmosphare except H_2/CO_2)

Reduction agent :
Na_2S and resazurin

23 July 2004 17

Batch Assay

Precursor substrates for methane production (H_2/CO_2, Formate und Methanol)

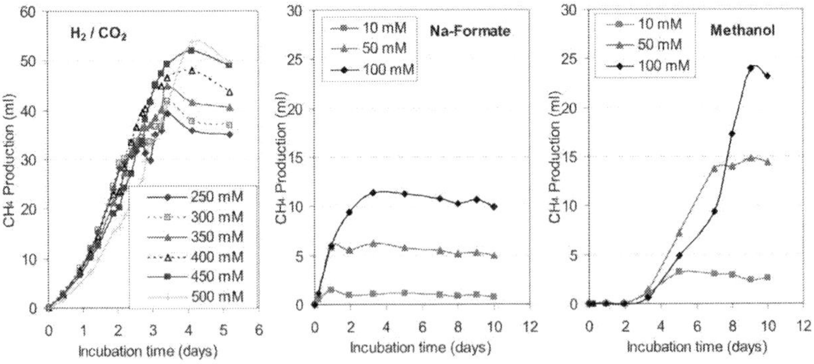

Increasing the concentration of substrate increased the methane production rate.

23 July 2004 18

Batch Assay

Na-Acetate as precursor substrate for methane production

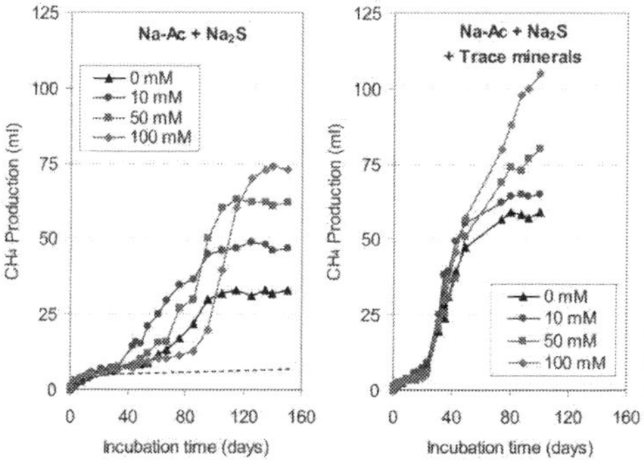

Substrate	V_{max} ($mmol \cdot l^{-1} \cdot hr^{-1}$)	K_m ($mmol \cdot l^{-1}$)
H_2 / CO_2	2.195	0.26
Na Formate	0.883	15
Methanol	0.600	30
Na Acetate		
1.Reduction agent	0.057	13
2. Reduction agent+ Trace minerals + Vitamins	0.065	27

Addition of a reduction agent (Na_2S) reduced the redox potentials → Methane production after the lag-Phase → Unsatisfied !

Addition of trace elements increased the V_{max}.

Suspended microorganisms in whey-reactor effluent

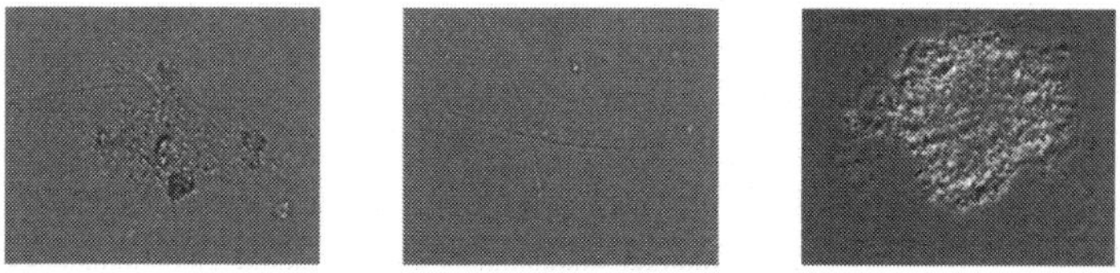

Total Bacterial Flora (Light Microscopy, 1000x)

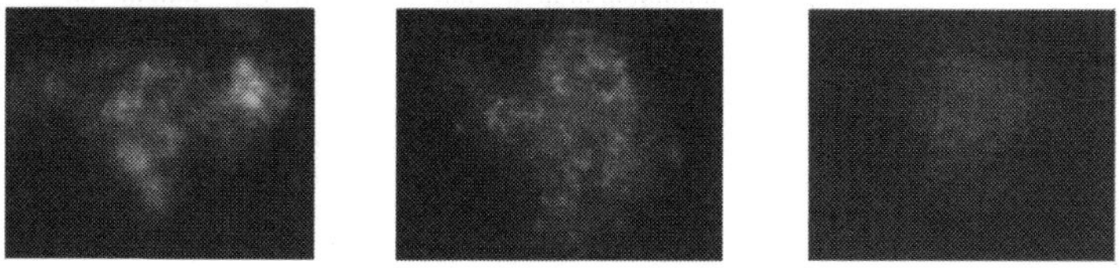

Methanogenic Bacteria (Fluorescent Microscopy, 1000x)

Effects of trace minerals on methane production and acetate degradation rates

Ni and Co are cofactors for enzyme (e. g. Hydrogenase).

A lack of these trace minerals can interupt the activity of anaerobic degradation.

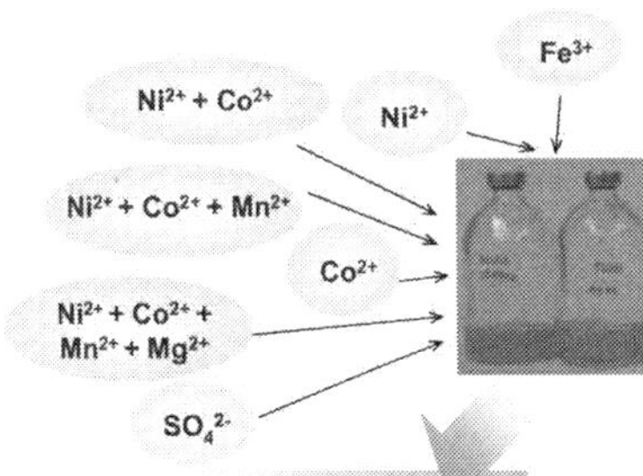

Condition:

- Suspended microorganism in whey-reactor effluent20% (vol/vol)
- In orbital shaker at 37°C
- Under anaeroben condition (N_2 Gas)

Substrate:
Residual fatty acids in the effluent of whey- reactor

Conversion rates:
- CH_3COOH conversion
- CH_4 Production

IMPLEMENTATION
Best result

Fixed Bed Reactor

Methane production from residual fatty acids in whey-reactor effluent with trace mineral additions

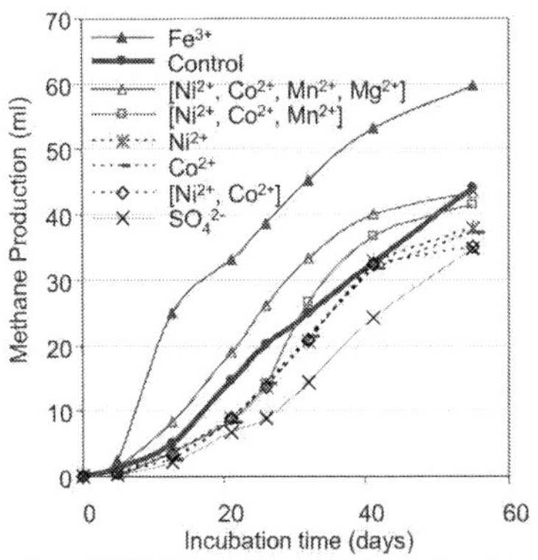

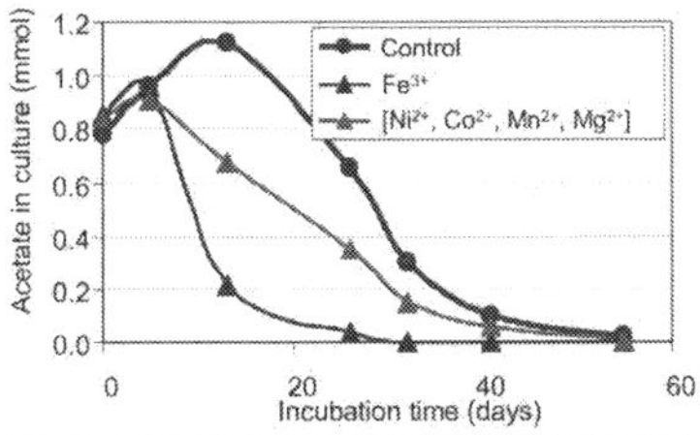

Variation	Rate ($mmol \cdot hr^{-1}$)	
	Acetate conversion	Methane production
control	-0.079	0.100
Fe^{3+}	-0.196	0.280
$[Ni^{2+}, Co^{2+}, Mn^{2+}, Mg^{2+}]$	-0.063	0.110

The addition of Fe^{3+} and a combination of Ni^{2+}, Co^{2+}, Mn^{2+} and Mg^{2+} increased the methane production and acetate degradation rates.

AnFB Reactor

Implementation by the addition of iron (III)

Continuous whey reactor systems

Effect of the concentration of Fe^{3+} on methane production in whey-reactor

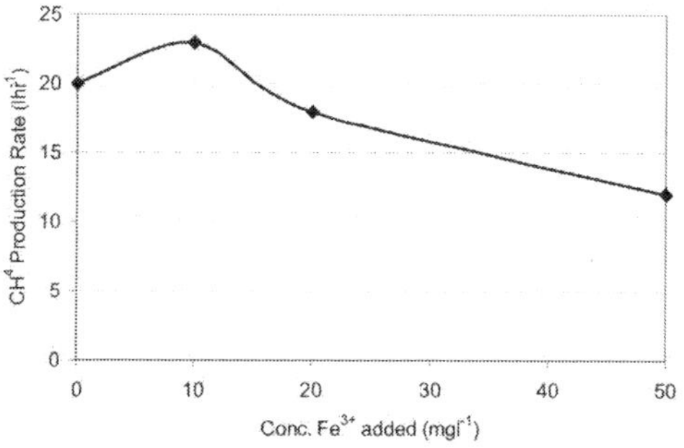

With an additional of 10 mg·l⁻¹ Fe^{3+}, the Methane production in the reactor was optimal.

23 July 2004 23

AnFB Reactor

Implementation the addition of iron (III)

Continuous whey reactor systems

Effect of the concentration of Fe^{3+} on acetate formation rates in whey-reactor

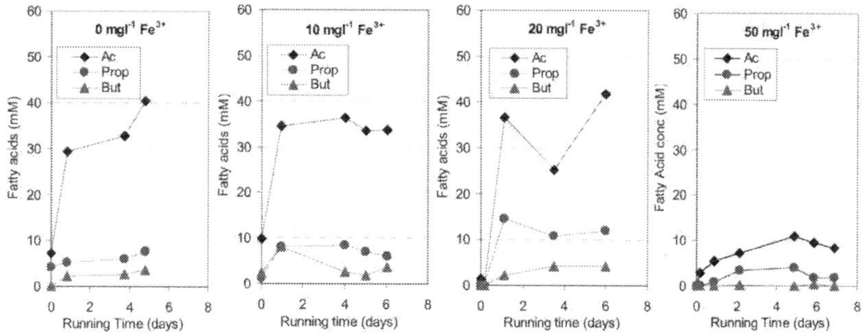

With an addition of 50 mg·l⁻¹ Fe^{3+}, the hydrolysis of whey was inhibited → less fatty acids produced → the methane production did not perform optimally.

23 July 2004 24

Conclusion

- AnFB reactor could be used for treating high concentration of whey-containing wastewater. Optimal operating conditions for whey treatment (Influent = 50 g $COD \cdot l^{-1}$) are:
  - A HRT of 4-8 days,
  - An OLR of ≤ 10 kg $COD \cdot m^{-3} \cdot d^{-1}$, and reach
  - 85% of COD removal

- Accumulation of fatty acids was triggered by surplus supply of whey solution at a high OLR or oxygenation. Conversion process (acetogenesis and methanogenesis) of fatty acids was influenced by the pH and concentration of fatty acids.

- The suspended methanogenic population in the AnFB contained:
  - Hydrogen-utilizing methanogenic bacteria,
  - Formate- and Methanol- utilizing methanogenic bacteria and
  - Acetoclastic methanogenic bacteria.

23 July 2004 25

- Acetate-utilizing methanogens, *Methanosarcina sp.* and *Methanosaeta sp.* preferred to grow in a particulate or attached manner on support material. The biofilm provides a low redox potential and a strict anaerobe environment. The growth of acetate-utilizing methanogens in suspended culture required an additional of reducing agent.

- H_2/CO_2 was the best methanogenic substrate for suspended bacteria followed by formate, methanol and acetate.

- Addition of Iron (III) or a combination of trace minerals improved acetate degradation and methane production more than two-folds.

- AnFB reactor is an alternative means for stabilizing wastewater from dairy industry. Liapor-clay-polyethylene sinter lamellas in a regularly arrangement could be the substratum for biofilm formation.

- In implementation, AnFB reactor should be operated at a minimum HRT of 4-6 days or a maximum OLR of 10 kg $COD \cdot m^{-3} \cdot d^{-1}$.
  This is concurrence with other anaerobic fixed bed reactors with different supporting media for dairy wastewater.

23 July 2004 26

Acknowledgement

- Prof. Dr. rer. nat. J. Winter
- Prof. Dr. Ing. E.h. H. H. Hahn, PhD.
- DFG-Graduiertenkolleg III GK-GRK 366/2-00 „Grenzflächenphänomene in aquatischen Systemen und wässrigen Phasen“ for the financial support

23 July 2004 27

Printed in Dunstable, United Kingdom

91335491R00087